ELECTRONICS
VEST POCKET
REFERENCE BOOK

ELECTRONICS VEST POCKET REFERENCE BOOK

by

Harry Thomas

PRENTICE-HALL, INC.
Englewood Cliffs, N. J.

Electronics Vest Pocket Reference Book by Harry Thomas

© 1969 by Harry Thomas

13-252361-2

Library of Congress Catalog Card Number: 70-82906
Printed in the United States of America · *T*

Prentice-Hall International, Inc., London
Prentice-Hall of Australia, Pty. Ltd., Sydney
Prentice-Hall of Canada, Ltd., Toronto
Prentice-Hall of India Private Ltd., New Delhi
Prentice-Hall of Japan, Inc., Tokyo

20 19 18 17 16 15 14

ACKNOWLEDGMENTS

Appreciation is expressed to the following companies for permission to use the indicated materials:

Chilton Publishing Co., Philadelphia, Pa., pages 144–45

Southern Screw Co., Statesville, N.C., page 185

Radio Corporation of America, Harrison, N.J., pages 118–20

Ohmite Manufacturing Co., Skokie, Ill., pages 52–53, 55–56

Fenwal Electronics Inc., Framingham, Mass., pages 206–7

Schweber Electronics, Westbury, N.Y., pages 208–15

General Radio Co., West Concord, Mass., pages 10–15

Hewlett-Packard, Palo Alto, Calif., pages 198–99

Weston Instruments, Inc., Newark, N.J., pages 173–174

CONTENTS

ELECTRONICS
VEST POCKET
REFERENCE BOOK

OHM'S LAW FOR DC CIRCUITS

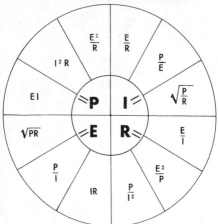

Fundamental Equations:

$$I = \frac{E}{R} = \frac{P}{E} = \sqrt{\frac{P}{R}}$$

$$R = \frac{E}{I} = \frac{P}{I^2} = \frac{E^2}{P}$$

$$E = IR = \frac{P}{I} = \sqrt{PR}$$

$$P = EI = \frac{E^2}{R} = I^2 R$$

Where: I = current in amperes
R = resistance in ohms
E = potential across R in volts
P = power in watts

Solution by Chart:

(1) Select P, E, I, R quadrant
(2) Read solution in adjacent sector

2

OHM'S LAW FOR AC CIRCUITS

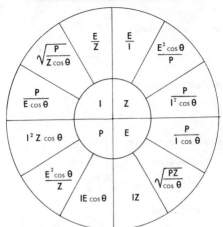

Fundamental Equations:

$$I = \frac{E}{Z} = \frac{P}{E \cos \theta} = \sqrt{\frac{P}{Z \cos \theta}}$$

$$Z = \frac{E}{I} = \frac{E^2 \cos \theta}{P} = \frac{P}{I^2 \cos \theta}$$

$$E = IZ = \frac{P}{I \cos \theta} = \sqrt{\frac{PZ}{\cos \theta}}$$

$$P = IE \cos \theta = I^2 Z \cos \theta = \frac{E^2 \cos \theta}{Z}$$

Where: I = current in amperes
Z = impedance in ohms
E = potential across Z in volts
P = power in watts
θ = phase angle in degrees

3

COMPONENT CIRCUIT FORMULAE

INDUCTIVE REACTANCE

$$X_L = 2\pi f L$$

RESISTORS IN SERIES

$$R_T = R_1 + R_2 + \cdots$$

RESISTORS IN PARALLEL

Two resistors:

$$R_T = \frac{R_1 R_2}{R_1 + R_2}$$

More than two:

$$\frac{1}{R_T} = \frac{1}{R_1} + \frac{1}{R_2} + \frac{1}{R_3} + \cdots$$

CAPACITIVE REACTANCE

$$X_C = \frac{1}{2\pi f C}$$

CAPACITORS IN SERIES

Two capacitors:

$$C_T = \frac{C_1 C_2}{C_1 + C_2}$$

More than two:

$$\frac{1}{C_T} = \frac{1}{C_1} + \frac{1}{C_2} + \frac{1}{C_3} + \cdots$$

CAPACITORS IN PARALLEL

$$C_T = C_1 + C_2 + \cdots$$

INDUCTORS IN PARALLEL

Two inductors:

$$L_T = \frac{L_1 L_2}{L_1 + L_2} \quad \text{(no coupling between coils)}$$

More than two:

$$\frac{1}{L_T} = \frac{1}{L_1} + \frac{1}{L_2} + \frac{1}{L_3} + \cdots$$
$$\text{(no coupling between coils)}$$

INDUCTORS IN SERIES

$$L_T = L_1 + L_2 + \cdots$$
$$\text{(no coupling between coils)}$$

IMPEDANCE IN AN RC CIRCUIT (SERIES)

$$Z = \sqrt{R^2 + X_C^2}$$

RL CIRCUIT TIME CONSTANT

$$\frac{L \text{ (in henrys)}}{R \text{ (in ohms)}} = t \text{ (in seconds)}$$

or

$$\frac{L \text{ (in microhenrys)}}{R \text{ (in ohms)}} = t \text{ (in microseconds)}$$

RC CIRCUIT TIME CONSTANT

R (ohms) $\times$ C (farads) $= t$ (seconds)
R (megohms) $\times$ C (microfarads) $= t$ (seconds)
R (ohms $\times$ C (microfarads) $= t$ (microseconds)
R (megohms) $\times$ C (micromicrofarads) $= t$ (microseconds)

AC CIRCUITS

RESISTIVE

$$\text{Impedance} = Z_R = R$$

Current is in phase with voltage.
Therefore, $\theta = 0$ and $\cos \theta = 1$

Power factor $= 1$

Power $= EI \cos \theta = EI$

Schematic

Waveform

INDUCTIVE

$$\text{Impedance} = X_L = 2\pi f L$$

Current lags voltage by 90°.
Therefore, $\theta = 90°$ and $\cos \theta = 0$

Power factor $= 0$

Power $= EI \cos \theta = 0$

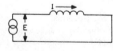

Schematic

Waveform

6

CAPACITIVE

$$\text{Impedance} = X_c = \frac{1}{2\pi f C}$$

Current leads voltage by 90°.
Therefore, $\theta = 90°$ and $\cos \theta = 0$

Power factor = 0

Power = $EI \cos \theta = 0$

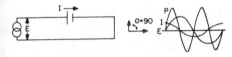

Schematic	Waveform

RESISTIVE AND INDUCTIVE (CAPACITIVE)

$$\text{Impedance} = Z = \sqrt{R^2 + X^2}$$

Current lags (leads) voltage by less than 90°.
Therefore, θ is variable = $\tan^{-1} X/R$

Power factor = $\cos \theta$

Power = $EI \cos \theta$

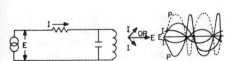

Schematic	Waveform

ALTERNATING CURRENT WAVEFORMS

1. SYMMETRICAL

Sinusoidal

Square

Non-sinusoidal

Triangular

2. NONSYMMETRICAL

Saw-tooth

3. SUPERIMPOSED ON DC

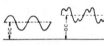

4. COMPOUNDED OF FUNDAMENTAL AND HARMONICS

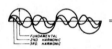 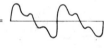

5. COMPOUNDED OF UNLIKE FREQUENCIES

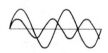 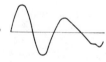

Sinusoidal Voltages and Currents

Effective value = 0.707 × peak value
Average value = 0.637 × peak value
Peak value = 1.414 × effective value
Effective value = 1.11 × average value
Peak value = 1.57 × average value
Average value = 0.9 × effective value

8

RESONANCE IN AC CIRCUITS

General:

Resonance occurs when the effects of inductance and capacitance exactly oppose each other. This occurs at a single particular frequency: $f = 1/(2\pi\sqrt{LC})$

Series Resonance:

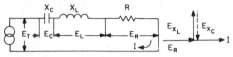

SERIES RESONANCE *occurs when:*

$$E_C = E_L \quad IX_C = IX_L \quad \text{or} \quad E_T = E_R = IR$$

AC voltages (and sometimes reactances) are added *vectorially.* Voltage across inductor (or inductive reactance) opposes voltage across capacitor (or capacitive reactance).

Series resonance causes voltage step-up. Voltage step-up is the ratio of the voltage across either of the reactive elements to the voltage across the source. Its value is $\dfrac{E_{XC}}{E_T} = \dfrac{E_{XL}}{E_T} = Q = \dfrac{X_C}{R} = \dfrac{X_L}{R}$

Impedance: $\qquad\qquad Z = R$

Parallel Resonance:

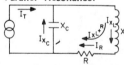

PARALLEL RESONANCE *occurs when:*

$$IX_C = IX_L \quad I_T = I_R$$

Branch currents are added *vectorially.* Current through inductor opposes capacitor current.

Parallel resonance causes current step-up. Current step-up is the ratio of the current through either of the reactive branches to the total line current. Its value is: $\dfrac{I_{XC}}{I_T} = \dfrac{I_{XL}}{I_T} = Q = \dfrac{X_C}{R} = \dfrac{X_L}{R}$

Impedance: $\qquad\qquad Z = \dfrac{1}{CR}$

REACTANCE CHARTS AND THEIR USAGE

The accompanying reactance charts provide a simple direct method for obtaining, at any frequency: (1) the reactance (in ohms) of any size inductance, (2) the reactance (in ohms) of any size capacitor, (3) the resonant frequency determined by any capacitor-inductance circuit combination.

Two charts, Fig. 1 and Fig. 2, are provided, Fig. 1 for rough calculations to obtain the *approximate range* of values needed and Fig. 2 for use as a magnified replica (7 : 1) of a single square area of Fig. 1 in order to obtain accurate values surrounding the point of interest.

Both charts have four sets of scale lines superimposed on one another. For explanation refer to Fig. 2.

One set of horizontal lines	Inductance scale	Extending *right to left* with number of ohms on left termination
One set of vertical lines	Frequency scale	Extending *from top to bottom* with numbers on bottom end of lines
One set of slant lines	Frequency scale	Extending *upward to right* from lower left to upper right. Numbered on outside border
One set of slant lines	Capacitance scale	Extending *downward to right* from upper left to lower right. Numbered on inside border

Correct use of the chart arrives at the desired electrical quantity by determining the intersection of two scale lines, each representing a given electrical quantity (inductance, capacitance, frequency, or reactance) and projecting this point along another line to the final value scaled on an edge of the chart.

After locating an intersection on the rough scale chart, Fig. 1, notice in which part or general area of a heavy-line-bounded-square the intersection lies; then go to the same general area of Fig. 2 (fine scale) since it is an expanded equivalent of the single rough-scale square where the intersection was found. This precaution is necessary since two levels of slant scale lines appear in each square.

EXAMPLE 1

Find reactance in ohms (see point + on rough scale) for 100 mmf at 720 kc

1. Locate intersection of 100-$\mu\mu$f capacitance line and 720-kc frequency line.

2. Draw horizontal line from intersection over to the left-hand reactance scale where it intersects the reactance scale at *2,200 ohms*—the answer.

3. On fine scale (Fig. 2) follow same procedure.

EXAMPLE 2

Find reactance of 500 microhenries (μh) at 720 kc

1. Locate intersection of 100-mmf capacitance line and 720-kc frequency line.

2. From this intersection draw horizontal line as above to 2,200-ohm point.

To determine the resonant frequency of a 100-$\mu\mu$f capacitor and a 500-μh coil select these two slant lines and locate their intersection at 720 kc.

11

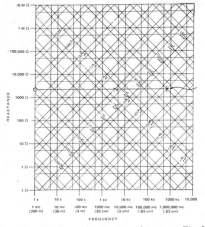

Figure 1. Use for approximation of entry on Fig. 2.

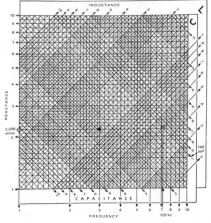

Figure 2. Use for obtaining exact value.

OHM'S LAW NOMOGRAPH

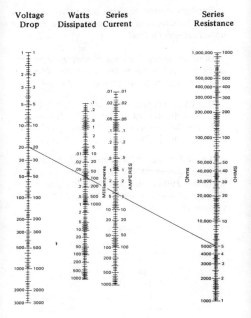

| Voltage Drop | Watts Dissipated | Series Current | | Series Resistance | |

Solve for unknown at intersecting straight edge between two given quantities.

Example: 20 volts across five ohms dissipates
80 watts

or

using Left Hand ohm, milliampere and watt scales:
20 volts across 5000 ohms
(4 milliamps) dissipates
80 milliwatts.

HOW TO CALCULATE DECIBELS

Decibels express nothing more than a power ratio. Calculations are based on the familiar powers-of-ten method.

In other words, for integral powers of ten the db value corresponds to ten times the exponent. For this conversion we take the following steps:

1. List power gain ratio, say   100 to 1
2. Rewrite exponentially 10^2 to 1
3. Multiply exponent by 10 db   2 X 10 db
4. Rewrite in final form   20 db

An easier way is to count the number of zeros or places to the right of the first digit, write this number down, and multiply it by ten. This procedure is summarized in the ratio conversion chart, Table 1.

Table 1.

Ratio		Decibels
1:	1	0
10:	1	10
100:	1	20
1,000:	1	30
10,000:	1	40
100,000:	1	50
1,000,000:	1	60
10,000,000:	1	70
100,000,000:	1	80

Extending this procedure, our system must be able to express intervening values in 1-db steps. For this no charts, books, tables, or slide rule are necessary. Only three ratios are used: 1.25, 1.6, and 2.0. These correspond respectively to 1, 2, and 3 db. Next, to use these less-than-10 db figures, we must remember that multiplying numbers is the same as adding exponents or the same as adding db's. Thus 10 (10 db) times 10,000 gives an answer equal to $10^{1+4} = 10^{+5} = 100,000$, or 50 db. Now using these smaller ratios to multiply 1.25 (1 db) by 1.6 (2 db) we see that the product (2) is equal to 3 db. Likewise 2 (3 db) times 2.5 (4 db) = 7 db. These db values between 1 and 10 plus their corresponding ratios are given in Table 2.

Table 2.

Ratio	Decibels
1.25 : 1	1
1.6 : 1	2
2.0 : 1	3
2.5 : 1	4
3.2 : 1	5
4.0 : 1	6
5.0 : 1	7
6.4 : 1	8
8.0 : 1	9
10.0 : 1	10

DECIMAL DB CALCULATIONS

For applications that require greater accuracy —that is, for tenth-db values—we start with the first of the basic numbers: 1.25, and derive a table in .1-db steps, by setting 1.25 at the top of the list, and subtracting from each succeeding value the amount in hundredths, as indicated by the following sequence.

$$1.0 \text{ db} = 1.25 - \quad 0 = 1.25 : 1$$
$$0.9 \text{ db} = 1.25 - .02 = 1.23 : 1$$
$$0.8 \text{ db} = 1.23 - .03 = 1.20 : 1$$
$$0.7 \text{ db} = 1.20 - .03 = 1.17 : 1$$
$$0.6 \text{ db} = 1.17 - .02 = 1.15 : 1$$
$$0.5 \text{ db} = 1.15 - .03 = 1.12 : 1$$

(Continued on next page)

DECIMAL DB CALCULATIONS (Cont.)

Table 3. Full list of tenths of db's.

Ratio	Decibels
1.25: 1	1.0
1.23: 1	0.9
1.20: 1	0.8
1.17: 1	0.7
1.15: 1	0.6
1.12: 1	0.5
1.09: 1	0.4
1.07: 1	0.3
1.04: 1	0.2
1.01: 1	0.1

EXAMPLE 1

A 400-watt amplifier with 1-watt input has a power ratio of 400: 1.

1. Breakdown: $400 = 100 \times 4$
   $= 10^2 \times 4$
2. Converting integral powers: $100 = 20$ db
3. Converting digital factors: $4 = 6$ db
4. Adding factor db's: $= 26$ db
   Total Gain

EXAMPLE 2

Find the power ratio that corresponds to 28.4 db.

Factor and convert: 20 db = 100 to 1 ratio
Convert digits: 8 db = 6.4 to 1 ratio
Convert decimals: .4 db = 1.09 to 1 ratio

(add)   (multiply)

28.4 db = 100×6.4
$\times 1.09 =$
697.6 to 1 ratio

DECIBEL—POWER CONVERSION

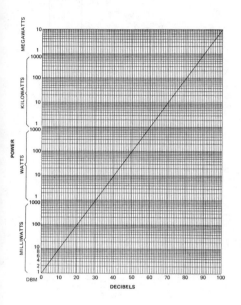

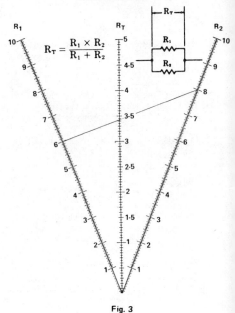

Fig. 3

RESISTANCES IN PARALLEL

Two-resistor Nomograph:

Solving for the equivalent value of two parallel resistors:

(1) Place straight edge (or draw a line) so that it intersects the two given values on the angularly diverging scales (R_1 and R_2).

(2) Desired equivalent resistance (Q_T) is at intersection on the vertical scale.

Example (Fig. 3)

(1) A line through:

$$R_1 = 6 \text{ ohms and } R_2 = 8 \text{ ohms}$$

(2) Intersects at

$$R_T = 3.42 \text{ ohms.}$$

Range extension can be effected multiplying the value of all scales by 10, 100, 1000 or as required.

For three resistances in parallel first find equivalent resistance of two of them and then solve for this one in parallel with the third.

SEMICONDUCTOR SYMBOLS AND ABBREVIATIONS

SYMBOL	MEANING
A	Ampere (AC rms or DC)
a	Ampere (peak)
B	Base electrode
BV	Breakdown voltage
BV_{COB}	Breakdown voltage, collector-to-base junction reverse biased (emitter open)
BV_{CEO}	Breakdown voltage, collector-to-emitter junction reverse biased (base open)
BV_{CER}	Breakdown voltage, collector-to-emitter junction reverse biased (specified resistance)
BV_{CES}	Breakdown voltage, collector-to-emitter junction reverse biased (base shorted)
BV_{CEX}	Breakdown voltage, collector-to-emitter junction reverse biased (base-emitter back biased)
C	Capacitance and collector electrode
C_e	Internal collector junction capacitance

SYMBOL	MEANING
C (dep)	Depletion layer capacitance
C (dif)	Diffusion capacitance
C_{ib}	Input capacitance, common base
I_C	Collector current (DC)
I_c	Collector current (AC rms)
i_c	Collector current (peak)
I_{CBO}, I_{CO}	Collector cutoff current (emitter open)
I_{CEO}	Collector cutoff current (base open)
I_{CER}	Collector cutoff current (specified resistor from base to emitter)
I_{CERV}	Collector cutoff current (reverse voltage on base)
I_{CES}	Collector cutoff current (base shorted to emitter)
I_E	Average emitter current (DC)
I_e	Instantaneous emitter current (AC rms)
i_e	Instantaneous emitter current (peak)
I_{EBO}	Emitter cutoff current (collector open)
I_F	Forward current (DC)
i_F	Forward current (peak)
i_{Fr}	Forward recovery current (specified instantaneous value)
I_o	Output current (DC)
I_R	Reverse current (DC)
i_R	Reverse current (peak)
i_{Rr}	Reverse recovery current (specified instantaneous value)
I_Z	Average Zener current
I_{ZK}	Zener knee current

SYMBOL	MEANING
I_{ZM}	Zener maximum current
I_{ZT}	Zener test current
i_Z	Instantaneous Zener current (AC rms)
K	Kilohm
L_c	Conversion loss
mA	Milliampere (average)
ma	Milliampere (peak)
mAa-c	Milliampere (AC rms)
mAd-c	Milliampere (DC)
μA	Microampere (average)
μa	Microampere (peak)
C_L	Load capacitance
C_{ob}	Output capacitance, common base
C_{oe}	Output capacitance, common emitter
E	Emitter electrode
f_{ab}	Common base small-signal short-circuit forward current transfer-ratio cutoff frequency (emitter input)
f_{ae}	Common emitter small-signal short-circuit forward current transfer-ratio cutoff frequency
f_{ref}	Reference frequency
f_{osc}	Maximum frequency of oscillation
f_{pg}	Power gain cutoff frequency
f_τ	Frequency at which common emitter current gain is unity
G_b	Power gain (common base)
G_c	Power gain (common collector)
G_e	Power gain (common emitter)
g_M, g_{FE}	Static or DC transconductance
g_m	Small-signal transconductance
G_M, G_{FE}	Large-signal transconductance
h_{fb}	Common base small-signal short-circuit forward current transfer-ratio

SEMICONDUCTOR SYMBOLS AND ABBREVIATIONS (Cont.)

SYMBOL	MEANING
h_{fe}	Common emitter small-signal short-circuit forward current transfer-ratio (base input)
h_{FE}	Common-emitter DC short-circuit forward current transfer-ratio current gain I_C/I_B
H_{FE}	Common emitter large-signal short-circuit forward current transfer-ratio
h_{ib}	Common base small-signal short-circuit input impedance (emitter input)
h_{ie}	Common emitter small-signal short-circuit input impedance (base input)
h_{ob}	Common base small-signal open-circuit output admittance (emitter input)
h_{oc}	Common collector small-signal open-circuit output admittance (base input)
h_{oe}	Common emitter small-signal open-circuit output admittance (base input)
h_{rb}	Common base small-signal open-circuit reverse-transfer voltage ratio (emitter input)
h_{rc}	Common collector small-signal open-circuit reverse-transfer voltage ratio (base input)
h_{re}	Common emitter small-signal open-circuit reverse-transfer voltage ratio (base input)
i_{surge}	Surge current
I_B	Average base current (DC)

SYMBOL	MEANING
I_b	Instantaneous base current (AC rms)
i_b	Base current (peak)
mW	Milliwatt (max, av, or rms)
mw	Milliwatt (peak)
NF	Noise figure
nsec	Millimicrosecond (nano)
Ω	Ohms
P	Power dissipation of all terminals (average total)
p	Power dissipation all terminals (peak)
P_b	Power dissipation of base, av
p_b	Power dissipation of base (peak)
P_c	Power dissipation of collector (average)
p_c	Power dissipation of collector (peak)
P_d	Power dissipation of device
P_e	Power dissipation of emitter (average)
p_e	Power dissipation of emitter (peak)
PIV	Peak inverse voltage
Q	Transistor
r'_b, r_{BB}'	Internal base spreading resistance
R_B	External base resistance
R_C	External collector resistance
R_E	External emitter resistance
R_{sat}	Saturation resistance
R_L	Resistance, load
T	Temperature
t	Time
T_A	Temperature, ambient
T_C	Temperature, case
TCBV	Temperature coefficient of breakdown voltage
t_d	Pulse delay time

SEMICONDUCTOR SYMBOLS AND ABBREVIATIONS (Cont.)

SYMBOL	MEANING
t_f	Pulse fall time
t_{fr}	Forward recovery time
T_J	Temperature junction
t_P	Pulse time
t_r	Pulse rise time
t_s	Pulse storage time
θ_{EA}	Thermal resistance, case to ambient
θ_{JA}	Thermal resistance, junction to air
θ_{JC}	Thermal resistance, junction to case
V	Volt (DC)
v	Volt (peak)
$V\text{a-c}$	Volt (AC rms)
V_{BB}	Base voltage (DC) supply
V_{BE}	Emitter voltage (DC) (base to emitter)
V_{CC}	Collector voltage (DC) supply
$V_{CE\,\text{sat}}$	Collector to emitter saturation voltage
V_{EC}	Emitter voltage (DC) (emitter to collector)
V_{EE}	Emitter voltage (DC) supply
V_{fe}	Floating potential
V_F	Forward voltage drop
V_o	Output voltage (DC)
v_o	Output voltage (AC)
V_{PT}	Voltage, punch-through
V_R	Reverse voltage (DC)
V_r	Reverse voltage (peak)
V_Z	Zener voltage
W	Watts (max, av, rms)
w	Watts (peak)
Z_z	Zener impedance
Z_{zk}	Zener impedance, knee
Z_{zt}	Zener impedance, test

ELECTRONIC CIRCUIT AND COMPONENT ABBREVIATIONS

SYMBOL	MEANING
AC	alternating current
a	ampere
A	area
AF	audio-frequency
C	capacitance
X_c	capacitive reactance
c	centi (10^{-2})
cm	centimeter
G	conductance
Q	coulomb
cemf	counterelectromotive force
I	current (DC or rms value)
i	current (instantaneous value)
d	deci (10^{-1})
dk	deka (10)
K, k	dielectric constant
E	difference in potential (DC or rms value)
e	difference in potential (instantaneous value)
DC	direct current
emf	electromotive force
f	frequency
G	giga (10^9)
h	hecto (10^2)
h	henry
hp	horsepower
Z	impedance
L	inductance
X_L	inductive reactance
k	kilo (10^3)
kv	kilovolt
kv-a	kilovolt-ampere
kw	kilowatt

ELECTRONIC CIRCUIT AND COMPONENT
ABBREVIATIONS (Cont.)

SYMBOL	MEANING
kw-hr	kilowatt-hour
H	magnetic field intensity
mmf.	magnetomotive force
M	mega (10^6)
$M\Omega$	megohm
μ	micro (10^{-6})
μa	microampere
μf	microfarad
μh	microhenry
$\mu\mu$f	micromicrofarad
μv	microvolt
m	milli (10^{-3})
ma	milliampere
mh	millihenry
mw	milliwatt
M	mutual inductance
n	nano (10^{-9})
Ω	ohm
p	pico (10^{-12})
pf	picofarad
P	power
R	resistance
rpm	revolutions per minute
rms.	root mean square
T	tera (10^{12})
t	time
T	torque
v	volt
w	watt

RF BANDS AND TRANSMISSION FREQUENCIES

BAND	FREQUENCY SPECTRUM (KC)	WAVELENGTH IN METERS
VLF (Very Low)	10—30	30,000—10,000
LF (Low)	30—300	10,000—1,000
MF (Medium)	300—3,000	1,000—100
HF (High)	3,000—30,000	100—10
VHF (Very High)	30,000—300,000	10—1
UHF (Ultra High)	300,000—3,000,000	1—1
SHF (Super High)	3,000,000—30,000,000	.1—.01
EHF (Extremely High)	30,000,000—300,000,000	.01—.001

RADAR BAND CODES AND FREQUENCIES

BAND P

FREQ	λ
0.225	133.3
0.390	76.9

BAND L

SUB	FREQ	λ
P	0.390 0.465	76.9 64.5
C	0.465 0.510	64.5 58.8

BAND S

SUB	FREQ	λ
E	1.55 1.65	19.3 18.2
F	1.65 1.85	18.2 16.2
T	1.85 2.00	16.2 15.0
C	2.00 2.40	15.0 12.5
Q	2.40 2.60	12.5 11.5

BAND X

SUB	FREQ	λ
A	5.20 5.50	5.77 5.45
Q	5.50 5.75	5.45 5.22
Y	5.75 6.20	5.22 4.84
D	6.20 6.25	4.84 4.80
B	6.25 6.90	4.80 4.35

BAND K

SUB	FREQ	λ
P	10.90 12.25	2.75 2.45
S	12.25 13.25	2.45 2.26
E	13.25 14.25	2.26 2.10
C	14.25 15.35	2.10 1.95
U	15.35 17.25	1.95 1.74

BAND Q

SUB	FREQ	λ
A	36.00 38.00	0.834 0.790
B	38.00 40.00	0.790 0.750
C	40.00 42.00	0.750 0.715
D	42.00 44.00	0.715 0.682
E	44.00 46.00	0.682 0.652

SUB	BAND V FREQ	λ
A	46.00 / 48.00	0.652 / 0.625
B	48.00 / 50.00	0.625 / 0.600
C	50.00 / 52.00	0.600 / 0.577
D	52.00 / 54.00	0.577 / 0.556
E	54.00 / 56.00	0.556 / 0.536

	FREQ	λ
T	17.25 / 20.50	1.74 / 1.46
Q	20.50 / 24.50	1.46 / 1.22
R	24.50 / 26.50	1.22 / 1.13
M	26.50 / 28.50	1.13 / 1.05
N	28.50 / 30.70	1.05 / 0.977
L	30.70 / 33.00	0.977 / 0.909
A	33.00 / 36.00	0.909 / 0.834

	FREQ	λ
R	6.90 / 7.00	4.35 / 4.29
C	7.00 / 8.50	4.29 / 3.53
L	8.50 / 9.00	3.53 / 3.33
S	9.00 / 9.60	3.33 / 3.13
X	9.60 / 10.00	3.13 / 3.00
F	10.00 / 10.25	3.00 / 2.93
K	10.25 / 10.90	2.93 / 2.75

	FREQ	λ
Y	2.60 / 2.70	11.5 / 11.1
G	2.70 / 2.90	11.1 / 10.3
S	2.90 / 3.10	10.3 / 9.68
A	3.10 / 3.40	9.68 / 8.83
W	3.40 / 3.70	8.83 / 8.11
H	3.70 / 3.90	8.11 / 7.69
Z	3.90 / 4.20	7.69 / 7.15
D	4.20 / 5.20	7.15 / 5.77

	FREQ	λ
L	0.510 / 0.725	58.8 / 41.4
Y	0.725 / 0.780	41.4 / 38.4
T	0.780 / 0.900	38.4 / 33.3
S	0.900 / 0.950	33.3 / 31.6
X	0.950 / 1.150	31.6 / 26.1
K	1.150 / 1.350	26.1 / 22.2
F	1.350 / 1.450	22.2 / 20.7
Z	1.450 / 1.550	20.7 / 19.3

MODULATION AND CARRIER WAVE RELATIONSHIPS

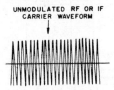

UNMODULATED RF OR IF
CARRIER WAVEFORM

Modulated Carrier Waveforms

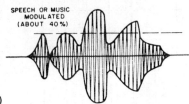

SPEECH OR MUSIC
MODULATED
(ABOUT 40 %)

(A)

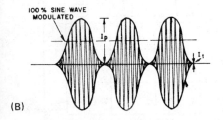

100 % SINE WAVE
MODULATED

I_p

I_t

(B)

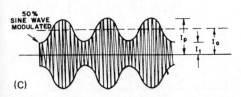

50 %
SINE WAVE
MODULATED

I_p

I_o

I_t

(C)

Percent Modulation—50% see (c)

$$M = \frac{I_p - I_t}{2 I_0} \times 100$$

$$= \frac{3 - 1}{4} \times 100$$

$$= 50\%$$

Modulating Audio Frequency Waveforms

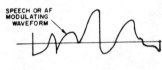

SPEECH OR AF
MODULATING
WAVEFORM

(D)

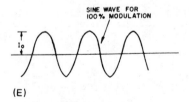

SINE WAVE FOR
100 % MODULATION

I_0

(E)

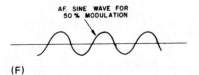

AF. SINE WAVE FOR
50 % MODULATION

(F)

SUMMARY OF SINE WAVE, FM AND PULSE MODULATION

TYPE OF MODULATION	MULTIPLEX SYSTEM	WAVEFORM	SYSTEM REQUIREMENTS	DISPLAY
AM Amplitude	Frequency division, analog		AM Detector, amplitude reference	Voltmeter, scope
FM Frequency	Frequency division, analog		FM Detector, frequency reference	Freq. meter, scope
PM Phase	Frequency division, analog		Phase detector, phase reference	Voltmeter
PAM Pulse amplitude	Time division, analog		AM Peak detector, amplitude ref., sync. pulses	Voltmeter, scope
PDM Pulse duration	Time division, analog		P.D. Detector, sync. pulses	Voltmeter, scope
PPM Pulse position	Time division, analog		Pulse interval detector, sync. pulses	Voltmeter, scope
PCM Pulse code (RZ) (Return to zero)	Time division, digital		Digital decoder, sync. pulses	Digital readout
PCM Pulse code (NZR) (Non-return to zero)	Time division, digital		Digital decoder, sync. pulses	Digital readout

TELEVISION CHANNEL FREQUENCIES

CHANNEL No.	Freq. Limits (MHz)	CHANNEL No.	Freq. Limits (MHz)	CHANNEL No.	Freq. Limits (MHz)
2	54–60	26	542–548	55	716–722
3	60–66	27	548–554	56	722–728
4	66–72	28	554–560	57	728–734
		29	560–566	58	734–740
5	76–82	30	566–572	59	740–746
6	82–88	31	572–578	60	746–752
		32	578–584	61	752–758
7	174–180	33	584–590	62	758–764
8	180–186	34	590–596	63	764–770
9	186–192	35	596–602	64	770–776
10	192–198	36	602–608	65	776–782
11	198–204	37	608–614	66	782–788
12	204–210	38	614–620	67	788–794
13	210–216	39	620–626	68	794–800
		40	626–632	69	800–806
14	470–476	41	632–638	70	806–812
15	476–482	42	638–644	71	812–818
16	482–488	43	644–650	72	818–824
17	488–494	44	650–656	73	824–830
18	494–500	45	656–662	74	830–836
19	500–506	46	662–668	75	836–842
20	506–512	47	668–674	76	842–848
21	512–518	48	674–680	77	848–854
22	518–524	49	680–686	78	854–860
23	524–530	50	686–692	79	860–866
24	530–536	51	692–698	80	866–872
25	536–542	52	698–704	81	872–878
		53	704–710	82	878–884
		54	710–716	83	884–890

TELEVISION CHANNEL CARRIER FREQUENCY LOCATIONS

Picture Carrier is always located:
   1.25 MHz *above* lower frequency limit.

Sound Carrier is always located:
   .25 MHz *below* upper frequency limit.

Difference between Picture and Sound Carriers is 4.5 MHz.

   Example:
   Channel 2: Lower Frequency Limit
               = 54 MHz
               Picture Carrier Frequency
               = 55.25 MHz

               Upper Frequency Limit
               = 60 MHz
               Sound Carrier Frequency
               = 59.75 MHz

INTERNATIONAL SYSTEM OF UNITS (SI)

Quantity	Unit		Symbol
Area	square meter	m^2	
Volume	cubic meter	m^3	
Frequency	hertz	Hz	(s^{-1})
Density	kilogram per cubic meter	kg/m^3	
Velocity	meter per second	m/s	
Angular velocity	radian per second	rad/s	
Acceleration	meter per second squared	m/s^2	
Angular acceleration	radian per second squared	rad/s^2	
Force	newton	N	$(kg \cdot m/s^2)$
Pressure	newton per square meter	N/m^2	
Kinematic viscosity	square meter per second	m^2/s	
Dynamic viscosity	newton-second per square meter	$N \cdot s/m^2$	
Work, energy, quantity of heat	joule	J	$(N \cdot m)$

(Continued on next page)

35

INTERNATIONAL SYSTEM OF UNITS (SI) (Cont.)

Quantity	Unit	Symbol	
Power	watt	W	(J/s)
Electric charge	coulomb	C	(A·s)
Voltage, potential difference, electromotive force	volt	V	(W/A)
Electric field strength	volt per meter	V/m	
Electric resistance	ohm	Ω	(V/A)
Electric capacitance	farad	F	(A·s/V)
Magnetic flux	weber	Wb	(V·s)
Inductance	henry	H	(V·s/A)
Magnetic flux density	tesla	T	(Wb/m²)
Magnetic field strength	ampere per meter	A/m	
Magnetomotive force	ampere	A	
Flux of light	lumen	lm	(cd·sr)
Luminance	candela per square meter	cd/m²	
Illumination	lux	lx	lm/m²

ENGLISH SYSTEM OF COMMON MEASUREMENTS

LINEAR

		1	inch	= 2.54 centimeters
1 foot	=	12	inches	= 30.48 centimeters
1 yard	=	3	feet	= 0.91 meter

1 rod = $\begin{cases} 5\frac{1}{2} \text{ yards} \\ 16\frac{1}{2} \text{ feet} \end{cases}$

1 statute mile = $\begin{cases} 320 \text{ rods} \\ 1760 \text{ yards} \\ 5280 \text{ feet} \end{cases}$ = 1609.34 meters

(1 furlong = ⅛ statute mile).

(1 internatioanl nautical mile = 1852 meters = 6076.11549 feet).

(The "knot" is a unit of speed, equal to 1 nautical mile per hour).

AREA

1 square foot =	144 square inches
1 square yard =	9 square feet
1 square rod =	$30\frac{1}{4}$ square yards

1 acre = $\begin{cases} 760 \text{ square rods} \\ 4,840 \text{ square yards} \\ 43,560 \text{ square feet} \end{cases}$ = 0.40 hectare

1 square mile =	640 acres	= 259.00 hectares

1 section of
of land = 1 mile square

CUBIC

1 cubic foot = 1728 cubic inches = 28.32 cubic decimeters

1 cubic yard = 27 cubic feet = 0.76 cubic meter

(Continued on next page)

ENGLISH SYSTEM OF COMMON MEASUREMENTS (Cont.)

LIQUID

1 fluid ounce =	8 fluid drams	=	29.57 milliliters
1 gill =	4 fluid ounces		
1 pint =	$\begin{cases} 4 \text{ gills} \\ 16 \text{ fluid ounces} \end{cases}$	=	473.16 milliliters
1 quart =	$\begin{cases} 2 \text{ pints} \\ 32 \text{ fluid ounces} \end{cases}$	=	0.95 liter
1 gallon =	$\begin{cases} 4 \text{ quarts} \\ 231 \text{ cubic inches} \end{cases}$	=	3.79 liters

DRY

1 quart =	2 pints	=	1.10 liters
1 peck =	8 quarts	=	8.81 liters
1 bushel =	$\begin{cases} 4 \text{ pecks} \\ 2150.42 \text{ cubic inches} \end{cases}$	=	3.52 dekaliters

ELECTRICAL ENGINEERING SYMBOLS AND UNITS

Quantity	Symbol	Unit	Symbol
Charge	Q	coulomb	C
Current	I	ampere	A
Voltage, potential difference	V	volt	V
Electromotive force	ξ	volt	V
Resistance	R	ohm	Ω
Conductance	G	mho (siemens)	A/V or mho (S)
Reactance	X	ohm	Ω

ELECTRICAL ENGINEERING SYMBOLS
AND UNITS (Cont.)

Quantity	Symbol	Unit	Symbol
Susceptance	B	mho	A/V or mho
Impedance	Z	ohm	Ω
Admittance	Y	mho	A/V or mho
Capacitance	C	farad	F
Inductance	L	henry	H
Energy, work	W	joule	J
Power	P	watt	W
Resistivity	ρ	ohm-meter	Ωm
Conductivity	σ	mho per meter	mho/m
Electric displacement	D	coulomb per sq. meter	C/m^2
Electric field strength	E	volt per meter	V/m
Permittivity (absolute)	ϵ	Farad per meter	F/m
relative permittivity	ϵ_r	(numeric)	
magnetic flux	Φ	weber	Wb
magnetomotive force	$\mathfrak{F}$	ampere (ampere-turn)	A
reluctance	$\mathfrak{R}$	ampere per weber	A/Wb
permeance	$\mathfrak{P}$	weber per ampere	Wb/A
magnetic flux density	B	tesla	T
magnetic field strength	H	ampere per meter	A/m
permeability (absolute)	μ	henry per meter	H/m
relative permeability	μ_r	(numeric)	
length	l	meter	m
mass	m	kilogram	kg
time	t	second	s
frequency	f	hertz	Hz
angular frequency	ω	radian per second	rad/s
force	F	newton	N
pressure	p	newton per sq. meter	N/m^2
temperature (absolute)	T	degree Kelvin	°K
temperature (International)	t	degree Celsius	°C

PHYSICAL AND ELECTRICAL CONVERSION FACTORS

To convert	Into	Multiply by	Conversely multiply by
Ampere-hours	Coulombs	3,600	2.778×10^{-4}
Amperes per sq. cm	Amperes per sq. inch	6.452	.155
Ampere turns	Gilberts	1.257	.7958
Ampere turns per cm	Ampere turns per inch	2.54	.3937
Btu (British thermal unit)	Foot-pounds	778.3	1.285×10^{-3}
Btu	Joules	1,054.8	9.48×10^{-4}
Btu	Kilogram-calories	.252	3.969
Btu	Horsepower-hours	3.929×10^{-4}	2,545
Centigrade	Fahrenheit	$(C° \times 9/5) + 32$	$(F° - 32) \times 5/9.$
Circular mils	Square centimeters	5.067×10^{-6}	1.973×10^{5}
Circular mils	Square mils	.7854	1.273.
Cubic inches	Cubic centimeters	16.39	6.102×10^{-2}
Cubic inches	Cubic feet	5.785×10^{-4}	1,728
Cubic inches	Cubic meters	1.639×10^{-5}	6.102×10^{4}
Cubic meters	Cubic feet	35.31	2.832×10^{-2}
Cubic meters	Cubic yards	1.308	.7646.
Degrees (angle)	Radians	1.745×10^{-2}	57.3
Dynes	Pounds	2.248×10^{-6}	4.448×10^{5}

Ergs	Foot-pounds	7.367×10^{-8}	1.356×10^7
Feet	Centimeters	30.48	3.281×10^{-2}
Foot-pounds	Horsepower-hours	5.05×10^{-7}	1.98×10^6
Foot-pounds	Kilogram-meters	.1383	7.233
Foot-pounds	Kilowatt-hours	3.766×10^{-7}	2.655×10^6
Gauss	Lines per sq. inch	6.452	.155
Grams	Dynes	980.7	1.02×10^{-3}
Grams	Ounces (avoirdupois)	3.527×10^{-2}	28.35
Grams per cm	Pounds per inch	5.6×10^{-3}	178.6
Grams per cubic cm	Pounds per cu. inch	3.613×10^{-2}	27.68
Grams per sq. cm	Pounds per sq. foot	2.0481	.4883
Horsepower (550 ft.-lb. per sec.)	Foot-lb. per minute	3.3×10^4	3.03×10^{-5}
Horsepower (550 ft.-lb. per sec.)	Btu per minute	42.41	2.357×10^{-2}
Horsepower (550 ft.-lb. per sec.)	Kg-calories per minute	10.69	9.355×10^{-2}
Horsepower (Metric) (542.5 ft.-lb. per sec.)	Horsepower (550 ft.-lb. per sec.)	.9863	1.014
Inches	Centimeters	2.54	.3937
Inches	Mils	1,000	.001
Joules	Foot-pounds	.7376	1.356.
Joules	Ergs	10^7	10^{-7}

(Continued on next page)

41

PHYSICAL AND ELECTRICAL CONVERSION FACTORS (Cont.)

To convert	Into	Multiply by	Conversely multiply by
Kilogram-calories	Kilojoules	4.186	.2389
Kilograms	Pounds (avoirdupois)	2.205	.4536
Kg per sq. meter	Pounds per sq. foot	.2048	4.882
Kilometers	Feet	3,281	3.048×10^{-4}
Kilometers	Btu	3,413	2.93×10^{-4}
Kilowatt-hours	Foot-pounds	2.655×10^{6}	3.766×10^{-7}
Kilowatt-hours	Joules	3.6×10^{6}	2.778×10^{-7}
Kilowatt-hours	Kilogram-calories	860	1.163×10^{-3}
Kilowatt-hours	Kilogram-meters	3.671×10^{5}	2.724×10^{-6}
Liters	Cubic meters	.001	1,000
Liters	Cubic inches	61.02	1.639×10^{-2}
Liters	Gallons (liq. US)	.2642	3.785
Liters	Pints (liq. US)	2.113	.4732
Meters	Yards	1.094	.9144
Meters per min	Feet per min	3.281	.3048
Meters per min	Kilometers per hr	.06	16.67
Miles (nautical)	Kilometers	1.853	.5396
Miles (statute)	Kilometers	1.609	.6214

Miles per hr	Kilometers per min	2.682×10^{-2}
Miles per hr	Feet per minute	88
Miles per hr	Kilometers per hr	1.609
Poundals	Dynes	1.383×10^4
Sq inches	Circular mils	1.273×10^6
Sq inches	Sq centimeters	6.452
Sq feet	Sq meters	9.29×10^{-2}
Sq miles	Sq yards	3.098×10^6
Sq miles	Sq kilometers	2.59
Sq millimeters	Circular mils	1,973
Tons, short (avoir 2,000 lb.)	Tonnes (1,000 Kg.)	.9072
Tons, long (avoir 2,240 lb.)	Tonnes (1,000 Kg.)	1.016
Tons, long (avoir 2,240 lb.)	Tons, short (avoir 2,000 lb)	1.120
Watts	Btu per min	5.689×10^{-2}
Watts	Ergs per sec	10^7
Watts	Ft-lb per minute	44.26
Watts	Horsepower (550 ft-lb per sec.)	1.341×10^{-3}
Watts	Horsepower (metric) (542.5 ft-lb per sec.)	1.36×10^{-3}
Watts	Kg-calories per min	1.433×10^{-2}

37.28
1.136×10^{-2}
.6214
7.233×10^{-5}
32.17
7.854×10^{-7}
.155
10.76
3.228×10^{-7}
.3861
5.067×10^{-4}
1.102
.9842
.8929
17.58
10^{-7}
2.26×10^{-2}
745.7
735.5
69.77

METRIC SYSTEM CONVERSION FACTORS

Length

Centimeter	=	0.3937 inch
Meter	=	3.28 feet
Meter	=	1.094 yards
Kilometer	=	0.621 statute mile
Kilometer	=	0.5400 nautical mile
Inch	=	2.54 centimeters
Foot	=	0.3048 meter
Yard	=	0.9144 meter
Statute mile	=	1.61 kilometers
Nautical mile	=	1.852 kilometers

Area

Sq centimeter	=	0.155 sq inch
Sq meter	=	10.76 sq feet
Sq meter	=	1.196 sq yards
Hectare	=	2.47 acres
Sq kilometer	=	0.386 sq mile
Sq inch	=	6.45 sq centimeters
Sq foot	=	0.0929 sq meter
Sq yard	=	0.836 sq meter
Acre	=	0.405 hectare
Sq mile	=	2.59 sq kilometers

Volume

Cu centimeter	=	0.0610 cu inch
Cu meter	=	35.3 cu feet
Cu meter	=	1.308 cu yards
Cu inch	=	16.39 cu centimeters
Cu foot	=	0.0283 cu meter
Cu yard	=	0.765 cu meter

Capacity

Milliliter	=	0.0338 U.S. fluid ounce
Liter	=	1.057 U.S. liq quarts
Liter	=	0.908 U.S. dry quart

U.S. fluid ounce	=	29.57 milliliters
U.S. liq quart	=	0.946 liter
U.S. dry quart	=	1.101 liters

Mass or Weight

Gram	=	15.43 grains
Gram	=	0.0353 avdp ounce
Kilogram	=	2.205 avdp pounds
Metric ton	=	1.102 short or net tons

Grain	=	0.0648 gram
Avdp ounce	=	28.35 grams
Avdp pound	=	0.4536 kilogram
Short or net ton	=	0.907 metric ton

Multiples and Submultiples	Prefixes	Symbols
$1,000,000,000,000 = 10^{12}$	tera	T
$1,000,000,000 = 10^{9}$	giga	G
$1,000,000 = 10^{6}$	mega	M
$1,000 = 10^{3}$	kilo	k
$100 = 10^{2}$	hecto	h
$10 = 10$	deka	dk
$0.1 = 10^{-1}$	deci	d
$0.01 = 10^{-2}$	centi	c
$0.001 = 10^{-3}$	milli	m
$0.000001 = 10^{-6}$	micro	μ*
$0.000000001 = 10^{-9}$	nano	n
$0.000000000001 = 10^{-12}$	pico	p

EXAMPLE: 1000 meters (or 10^3 meters) is called a kilometer, and one millionth of a gram (or 10^{-6} gram) is called a microgram.

*1 millionth of a meter is called a *micron*, and is abbreviated simply μ.

45

TEMPERATURE CONVERSION TABLES

FAHRENHEIT TO CENTIGRADE*

For positive Centigrade temperature, read *downward* from 0.00°C.

For negative Centigrade temperature, *read upward* from 0.00°C.

Temperature Fahrenheit	0	1	2	3	4	5	6	7	8	9
100	73.33	73.89	74.44	75.00	75.56	76.11	76.67	77.22	77.78	78.33
90	67.78	68.33	68.89	69.44	70.00	70.56	71.11	71.67	72.22	72.78
80	62.22	62.78	63.33	63.89	64.44	65.00	65.56	66.11	66.67	67.22
70	56.67	57.22	57.78	58.33	58.89	59.44	60.00	60.56	61.11	61.67
60	51.11	51.67	52.22	52.78	53.33	53.89	54.44	55.00	55.56	56.11
50	45.56	46.11	46.67	47.22	47.78	48.33	48.89	49.44	50.00	50.56
40	40.00	40.56	41.11	41.67	42.22	42.78	43.33	43.89	44.44	45.00
30	34.44	35.00	35.56	36.11	36.67	37.22	37.78	38.33	38.89	39.44
20	28.89	29.44	30.00	30.56	31.11	31.67	32.22	32.78	33.33	33.89
10	23.33	23.89	24.44	25.00	25.56	26.11	26.67	27.22	27.78	28.33
0	17.78	18.33	18.89	19.44	20.00	20.56	21.11	21.67	22.22	22.78
0	17.78	17.22	16.67	16.11	15.56	15.00	14.44	13.89	13.33	12.78
10	12.22	11.67	11.11	10.56	10.00	9.44	8.89	8.33	7.78	7.22
20	6.67	6.11	5.56	5.00	4.44	3.89	3.33	2.78	2.22	1.67
30	1.11	0.56	0.00	0.56	1.11	1.67	2.22	2.78	3.33	3.89
40	4.44	5.00	5.56	6.11	6.67	7.22	7.78	8.33	8.89	9.44
50	10.00	10.56	11.11	11.67	12.22	12.78	13.33	13.89	14.44	15.00
60	15.56	16.11	16.67	17.22	17.78	18.33	18.89	19.44	20.00	20.56
70	21.11	21.67	22.22	22.78	23.33	23.89	24.44	25.00	25.56	26.11
80	26.67	27.22	27.78	28.33	28.89	29.44	30.00	30.56	31.11	31.67
90	32.22	32.78	33.33	33.89	34.44	35.00	35.56	36.11	36.67	37.22
100	37.78	38.33	38.89	39.44	40.00	40.56	41.11	41.67	42.22	42.78

*Basic formula: $°C = (5/9)(°F - 32)$

CENTIGRADE TO FAHRENHEIT*

Temperature Centigrade	0	1	2	3	4	5	6	7	8	9
100	148.0	149.8	151.6	153.4	155.2	157.0	158.8	160.6	162.4	164.2
90	130.0	131.8	133.6	135.4	137.2	139.0	140.8	142.6	144.4	146.2
80	112.0	113.8	115.6	117.4	119.2	121.0	122.8	124.6	126.4	128.2
70	94.0	95.8	97.6	99.4	101.2	103.0	104.8	106.6	108.4	110.2
60	76.0	77.8	79.6	81.4	83.2	85.0	86.8	88.6	90.4	92.2
50	58.0	59.8	61.6	63.4	65.2	67.0	68.8	70.6	72.4	74.2
40	40.0	41.8	43.6	45.4	47.2	49.0	50.8	52.6	54.4	56.2
30	22.0	23.8	25.6	27.4	29.2	31.0	32.8	34.6	36.4	38.2
20	4.0	5.8	7.6	9.4	11.2	13.0	14.8	16.6	18.4	20.2
10	14.0	12.2	10.4	8.6	6.8	5.0	3.2	1.4	0.4	2.2
0	32.0	30.2	28.4	26.6	24.8	23.0	21.2	19.4	17.6	15.8
0	32.0	33.8	35.6	37.4	39.2	41.0	42.8	44.6	46.4	48.2
10	50.0	51.8	53.6	55.4	57.2	59.0	60.8	62.6	64.4	66.2
20	68.0	69.8	71.6	73.4	75.2	77.0	78.8	80.6	82.4	84.2
30	86.0	87.8	89.6	91.4	93.2	95.0	96.8	98.6	100.4	102.2
40	104.0	105.8	107.6	109.4	111.2	113.0	114.8	116.6	118.4	120.2
50	122.0	123.8	125.6	127.4	129.2	131.0	132.8	134.6	136.4	138.2
60	140.0	141.8	143.6	145.4	147.2	149.0	150.8	152.6	154.4	156.2
70	158.0	159.8	161.6	163.4	165.2	167.0	168.8	170.6	172.4	174.2
80	176.0	177.8	179.6	181.4	183.2	185.0	186.8	188.6	190.4	192.2
90	194.0	195.8	197.6	199.4	201.2	203.0	204.8	206.6	208.4	210.2
100	212.0	213.8	215.6	217.4	219.2	221.0	222.8	224.6	226.4	228.2

For negative Fahrenheit temperature, *read upward from 32°F.*

For positive Fahrenheit temperature, *read downward from 32°F.*

*Basic formula: °F = (9/5)°C + 32

47

TEMPERATURE COEFFICIENT OF RESISTANCE

The resistance alloys used for all except the lowest ohmic values show such little change with temperature that in most power circuits the resistance is considered constant. Actually there may be changes at *full load* of -4% to $+8\%$ of the initial resistance. The change is usually referred to in terms of the "temperature coefficient of resistance" which is the change in "ohms-per-ohm-per-degree change in temperature" of the wire.

For special applications which require very constant resistance, it becomes desirable to specify the maximum permissible TC (temperature coefficient of resistance) and the range of temperature; and consequently to use only certain types of resistance alloys. The presently known low TC alloys, in the 800 ohms per circular milfoot class, consist largely of nickel and chromium with small amounts of aluminum and either copper or iron. The low-resistance alloys, 294 ohms per circular milfoot, consist largely of nickel and copper with traces of other metals.

Both of these classes of wire are rated by the wire manufacturers as having a TC of $\pm.00002$ ohm/ohm/°C (also called .002% or 20 parts per million) over a *limited range*, generally $-55°$ to $+150°$C and $+20°$ to $+100°$C, respectively; compared to conventional $+170$ PPM.

Unfortunately, the TC of a completed power resistor is generally different from that of the original wire, as the TC is affected by the materials, heat treatment, methods of construction and the increased temperature range. Without special control, the TC over the range to 300°C rise may increase to as much as ±80 PPM from the original ±20 PPM. Smaller TC's increase the cost of manufacture. Theoretical changes in resistance with temperature are shown in the accompanying figure.

The circuit designer should carefully consider the actual needs of the circuit before specifying limits on the TC of a desired resistor. Wherever possible it is best to select a critical resistor so as to operate at a low temperature rise. This will also provide the maximum stability over a long period.

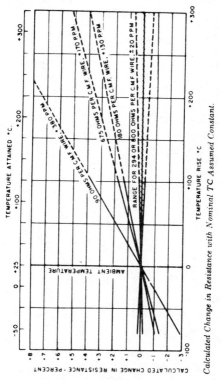

Calculated Change in Resistance with Nominal TC Assumed Constant.

49

SPECIFIC RESISTANCE OF CONDUCTING METALS

Type of metal	Ohms per circular mil-foot at 20°C
Copper (annealed)	10.35
Copper (hard drawn)	10.60
Aluminum	17.
Brass	42.
German silver (18%)	198.
Gold	14.6
Iron (pure, annealed)	61.
Iron (cast)	435.
Lead	132.35
Manganin	264.
Mercury	576.
Molybdenum	34.
Monel metal	252.
Nichrome	675.
Nickel	60.
Platinum	60.
Silver	9.56
Steel (soft, carbon)	96.
Tantalum	93.
Tin	69.
Tungsten	34.
Zinc	35.

HOW TO CALCULATE TEMPERATURE RISE FROM RESISTANCE MEASUREMENTS

Heat rise (degrees centigrade) in a transformer or copper wire relay winding can be determined accurately by winding-resistance measurement and calculation:

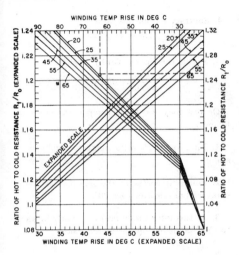

Two scales are presented on chart for greater convenience. Numbers on curves refer to initial winding temperature in degrees C.

1. Measure the winding resistance cold (at room temperature, assumed to be 20°); call it R_0 (say, 16 ohms).

2. Measure the final resistance at the end of the heat run; call this R_t (say, 20 ohms).

(Continued on next page)

HOW TO CALCULATE
TEMPERATURE RISE FROM
RESISTANCE MEASUREMENTS (Cont.)

3. Calculate the resistance ratio of the hot winding to that of the winding cold:

$$\frac{R_t}{R_0} = \frac{20}{16} = 1.25$$

4. Subtract 1 from this ratio:

$$(1.25 - 1) = .25.$$

5. Divide this figure (.25) by .00393:

$$\frac{.25}{.00393} = 63.20°.$$

Or one may use the accompanying chart as follows:

1. Find the 1.25 ratio point (on right-hand scale).

2. Follow it across to the slanting *cold* temperature line—in this case 20°.

3. At this intersection project directly upward to 63.2° point on top horizontal scale.

ELECTRONIC CONSTANTS, MULTIPLES AND SUB-MULTIPLES

Constant	Symbol	Rounded Value
electronic charge	e	1.602×10^{-19} C
speed of light in vacuum	c	2.9979×10^{8} m/s
permittivity of vacuum, electric constant	$\epsilon_o, \ \Gamma_e$	8.8542×10^{-12} F/m
permeability of vacuum, magnetic constant	$\mu_o, \ \Gamma_m$	$4\pi \times 10^{-7}$ H/m†
Planck constant	h	6.63×10^{-34} Js
Boltzmann constant	k	1.38×10^{-23} J/°K
Faraday constant	F	9.649×10^{4} C/mol
proton gyromagnetic ratio	γ	2.6752×10^{8} rad/sT
standard gravitational acceleration	g_n	9.80665 m/s²†
normal atmospheric pressure	atm	$101\ 325$ N/m²†

†Defined value.

Electrical Multiples and Submultiples

Multiply reading in	By	To obtain reading in
Amperes	1,000,000,000,000	micromicro-amperes
Amperes	1,000,000	microamperes
Amperes	1,000	milliamperes
Cycles	.000,001	megacycles
Cycles	.001	kilocycles
Farads	1,000,000,000,000	micromicro-farads
Farads	1,000,000	microfarads
Farads	1,000	millifarads
Henrys	1,000,000	microhenrys
Henrys	1,000	millihenrys
Kilocycles	1,000	cycles
Kilowatts	1,000	watts
Megacycles	1,000,000	cycles
Mhos	1,000,000	micromhos
Mhos	1,000	millimhos
Microamperes	.000,001	amperes
Microfarads	.000,001	farads
Microhenrys	.000,001	henrys
Micromhos	.000,001	mhos
Microvolts	.000,001	volts
Micromicro-farads	.000,000,000,001	farads
Milliamperes	.001	amperes
Millihenrys	.001	henrys
Millimhos	.001	mhos
Millivolts	.001	volts
Milliwatts	.001	watts
Volts	1,000,000	microvolts
Volts	1,000	millivolts
Watts	1,000	milliwatts

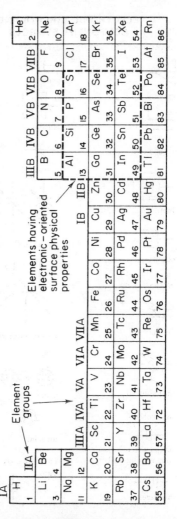

ELEMENT GROUPING IN PERIODIC ATOMIC TABULATION

Element groups

Elements having electronic – oriented surface physical properties

IA	IIA	IIIA	IVA	VA	VIA	VIIA			IB	IIB	IIIB	IVB	VB	VIB	VIIB	
H 1																He 2
Li 3	Be 4										B 5	C 6	N 7	O 8	F 9	Ne 10
Na 11	Mg 12										Al 13	Si 14	P 15	S 16	Cl 17	Ar 18
K 19	Ca 20	Sc 21	Ti 22	V 23	Cr 24	Mn 25	Fe 26	Co 27	Ni 28	Cu 29	Zn 30	Ga 31	Ge 32	As 33	Se 34	Br 35	Kr 36
Rb 37	Sr 38	Y 39	Zr 40	Nb 41	Mo 42	Tc 43	Ru 44	Rh 45	Pd 46	Ag 47	Cd 48	In 49	Sn 50	Sb 51	Te 52	I 53	Xe 54
Cs 55	Ba 56	La 57	Hf 72	Ta 73	W 74	Re 75	Os 76	Ir 77	Pt 78	Au 79	Hg 80	Tl 81	Pb 82	Bi 83	Po 84	At 85	Rn 86

55

ELECTRONIC CIRCUIT AND COMPONENT SYMBOLS

ANTENNAS

general

dipole

loop

counterpoise

ATTENUATORS

balanced

unbalanced

BATTERIES

DC source;
general—one cell

multicell

CAPACITOR

general

shielded

adjustable

mechanically linked

variable differential

CORE

memory

toroidal

**CRYSTAL UNIT, PIEZOELEC-
TRIC**

EARPHONE, TELEPHONE RECEIVER, HEARING AID

general

FUSE

general

fusible element

INDUCTOR WINDING

general

magnetic core

tapped

adjustable

continuously adjustable

LAMPS

ballast lamp or tube

AC glow, cold cathode or neon

incandescent

MICROPHONE

general

directional

PICKUP

mechanoelectric (complete)

RECTIFIER, METALLIC

general

full-wave bridge

ELECTRONIC CIRCUIT AND
COMPONENT SYMBOLS (Cont.)

RELAY

basic

coil

2-pole, double make

polarized relay with
transfer contact

RESISTOR

general

tapped

adjustable contact
variable

SHIELDING

electric or magnetic

SIGNALLING DEVICES

bell

buzzer

horn, loudspeaker,
siren, howler

sounder, telegraph

SWITCH

general–single throw

general–double throw

general–knife

pushbutton–circuit
closing (make)

pushbutton–circuit
opening (break)

SYNCHRO

may have letter symbol
to indicate type

transformer, receiver,
transmitter (complete)

differential receiver,
transmitter (complete)

TELEGRAPH KEY

(complete)

TERMINAL BOARD OR STRIP

(4 terminals shown)

THERMISTOR

general (complete)

with independent
integral heater (complete)

THERMOCOUPLE
(all symbols complete)

general

TRANSFORMER

general; in coaxial and
waveguide, taper or step
transformer without mode
change

magnetic core

shielded

adjustable inductance–
1 winding adjustable

mutual inductor,
adjustable inductance–
constant current trans-
former

autotransformer

59

ELECTRONIC CIRCUIT AND
COMPONENT SYMBOLS (cont'd.)

VARISTOR

asymmetrical

symmetrical

VIBRATOR

shunt drive (typical)

separate drive (typical)

SEMICONDUCTOR DEVICES

NPN transistor

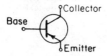

PNP transistor

Unijunction
transistor (UJT)

SEMICONDUCTOR DEVICES
(Continued)

Field effect
transistor (FET)

Diode Rectifier

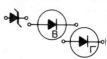

Zener (avalanche) diode

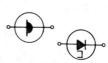

Tunnel diode

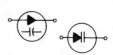

Varactor diode

Light sensor diode

Light emissive diodes
(injection lasers)

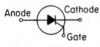

Silicon controlled
rectifier (SCR)

Silicon controlled
switch (SCS)

Gate turn-off
switch (GTO)

Triac (bidirectional
triode)

LASCR (light
activated SCR)

LASCS (light
activated switch)

TUBE, ELECTRON

triode

pentode, equipotential
cathode

twin triode, equipotential
cathode

rectifier; DC voltage
regulator (DC glow lamp
is same)

phototube

photomultiplier

61

ELECTRONIC CIRCUIT AND
COMPONENT SYMBOLS (cont'd.)

TUBE, ELECTRON
(Continued)

vapor rectifier, single-
anode, pool type with
ignitor

magnetron, tunable,
aperture coupled

reflex klystron, integral
cavity, aperture coupled

COAXIAL CABLE

(used with recognition
symbols at path ends and
intermediate points)

ATTENUATOR, UNIDIREC-
TIONAL

(shown: attenuation in
main direction, 1 db; opposite direction, 55 db)

CIRCULATOR

(arrow as shown indicates
power flow to next arm,
but not to others)

COUPLER, DIRECTIONAL

general; arrows indicate
direction of power flow

aperture coupling, 30 db
loss

loop coupling

probe coupling

resistance coupling

COUPLING

by loop to space

by loop to guided trans-
mission path

by probe to space

COUPLING (Continued)

by loop from coaxial to circular waveguide with DC grounds connected

by probe from co-axial to rectangular waveguide with DC grounds connected

FIELD POLARIZATION

rotator

amplitude modulator

FLANGES

→

plain

—≺

choke

HYBRID

general

rectangular waveguide and coaxial coupling

ISOLATION, DC

—✕—

of waveguide path (intentional)

RESONATOR

C

general

tunable, with adjustable Q coupled by probe to coaxial system

TERMINATORS

—◁

cable (line at left)

—|

open circuit (not a fault)

short circuit (not a fault)

coaxial cable and waveguide—movable short

—)|—

series capacitor, path open

series inductor, path shorted

TRANSDUCER, MODE

general

from rectangular to circular waveguide

WAVEGUIDE

—○—

circular

—|—

rectangular

ridged

GENERALIZED ELECTRONIC COMPONENT BREAKDOWN

	Symbol	Designation	Type and construction	Electronic application		Physical nature
Resistors		R	Composition Wirewound Film	Dissipative		Homogeneous conductors
Capacitors		C	Paper-air Mica-glass Deposited film Ceramic Electrolytic film	Storage	Passive	Dielectric insulated Plates
Inductors		L	Coiled conductor with or without ferrous core Audio, P.S. or high frequency	Storage		Self-inductive impedance
Vacuum tubes		V	Light emissive Photo-generative Receiver Transmitter	Amplifier Detector Oscillator Display Photo-transducer	Active	Electrodes in vacuum

Transistors		Q	Silicon Germanium Low signal Power circuits Computer	Active	Amplifier Detector Oscillator Switching Photo-transducer	Solid-state double junctions
Diodes		CR	Germanium-silicon Low signal Power Computer Tunnel Depletion layer Photo-emmissive	Passive	Detector Power rectifier Power switching Signal switching Parametric effect	Solid-state single junction
Crystal diodes		CR	Piezo-electric Quartz Rochelle salt	Active	Oscillator Frequency standard Physical transducer	Metal plate crystal sandwich

COMMON VACUUM TUBE SYMBOLS

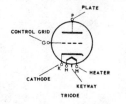

TRIODE

DIODE

CONVENTIONAL TUBES

DUO - DIODE

DIODE - TRIODE

DIODE - BEAM POWER

DIODE - PENTODE   DUO - TRIODE

DUO-DIODE-TRIODE

DUO-DIODE-PENTODE

TRIPLE - DIODE-TRIODE

DUO - PENTODE   TRIODE - PENTODE

DIODE - TRIODE-PENTODE

TRIODE - HEXODE

TRIODE - HEPTODE

THYRATRON

V-R TUBE

GAS FILLED TUBES

66

SEMICONDUCTOR LETTER SYMBOLS

ROMAN LETTERS

A_f	Band pass of a resonant circuit
agc	Automatic gain control
A_i	Current gain
am	Amplitude modulation
A_v	Voltage gain
bfo	Beat frequency oscillator
C	Constant (when following an equal ($=$) sign)
CB	Common-base configuration
C_{cb}	Interelement capacitance between base and collector
C_{eb}	Interelement capacitance between base and emitter
C_{cc}	Interelement capacitance between collector and emitter
CC	Common-collector configuration
CE	Common-emitter configuration
cm	Centimeter
C_N	Neutralizing capacitor
CR	Junction diode
C_{sp}	Reflected capacitance
cw	Continuous-wave transmission
E_{IN}	Input voltage
E_{OUT}	Output voltage
fm	Frequency modulation
f_a	Cutoff frequency
f_r	Frequency of resonance
G	Power gain
g_{bc}	Base-collector conductance
g_{be}	Base-emitter conductance
g_{ce}	Collector-emittance conductance
g_{fe}	Forward transfer conductance

g_{ie}	Input conductance
G_m	Maximum power
g_{oe}	Output conductance
g_{re}	Reverse transfer conductance
h	Hybrid
h_{ib}	Input resistance of common-base configuration with output short-circuited
h_{ic}	Input resistance of common-collector configuration with output short-circuited
h_{ie}	Input resistance of common-emitter configuration with output short-circuited
h_{ob}	Output conductance of common-base configuration with input open
h_{oc}	Output conductance of common-collector configuration with input open
h_{oe}	Output conductance of common-emitter configuration with input open
I_e	Total emitter current
I_b	Changing base current (AC)
I_B	DC base current
i_c	Changing collector current (AC)
I_C	DC collector current
if	Intermediate frequency
m	Turns ratio of a transformer
mA	Milliampere
mV	Millivolt
mW	Milliwatt
N	Number of turns in an inductor

N-type	Semiconductor with donor impurity
NPN	Transistor with one P-type and two N-type semiconductors
P-type	Semiconductor with acceptor impurity
P	Power
PN	Combination of N-type and P-type semiconductors
PNP	Transistor with one N-type and two P-type semiconductors
Q	Selectivity of a resonant circuit
RT	Thermistor
r_b	AC base resistance
RC	A coupling circuit employing a resistor and a capacitor
r_d	AC collector resistance
r_e	AC emitter resistance
r_{fe}	Forward transfer resistance with input open
r_{ie}	Input resistance with output open
R_L	Load resistor
r_m	Mutual resistance
R_N	Neutralizing resistor
r_{oe}	Output resistance with input open
r_{re}	Reverse transfer resistance with input open
S_I	Current-stability factor
S_V	Voltage-stability factor
T	Transformer
TC	Time constant
t_f	Fall time
t_p	Pulse time
t_r	Rise time
t_s	Storage time
V_{BE}	Fixed base-emitter voltage
V_{CE}	Fixed collector-emitter voltage

SEMICONDUCTOR LETTER
SYMBOLS (Cont.)

V_g	Source voltage
v_{be}	Changing voltage (AC) between base and emitter
v_{ce}	Changing voltage (AC) between collector and emitter
V_{SAT}	Saturation voltage
Y	Admittance
Z	Impedance

GREEK LETTERS

α	Alpha; current amplification factor
α_{fb}	Forward short-circuit current amplification factor for the CB configuration
α_{fc}	Forward short-circuit current amplification factor for the CC configuration
α_{fe}	Forward short-circuit current amplification factor for the CE configuration
Δ	Delta; incremental change
∞	Infinity
μA	Microampere
μ_{rb}	Reverse open-circuited voltage amplification factor for CB configuration
μ_{rc}	Reverse open-circuited voltage amplification factor for CC configuration
μ_{re}	Reverse open-circuited voltage amplification factor for CE configuration

AXIAL AND RADIAL LEAD
RESISTOR MARKINGS

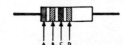

70

axial leads	color	radial leads
Band A	indicates first significant figure of resistance value in ohms.	Body A
Band B	indicates second significant figure.	End B
Band C	indicates decimal multiplier.	Band C or dot
Band D	if any, indicates tolerance in percent about nominal resistance value. If no color appears in this position. tolerance is 20%.	Band D

Note: Low-power insulated wire-wound resistors have axial leads and are color coded similar to the left-hand figure at bottom of facing page, except that band A is double width.

RESISTANCE AND CAPITANCE NUMBER CODING

COLOR	SIGNIFICANT FIGURE OR NUMBER OF ZEROS	DECIMAL MULTIPLIER	RESISTANCE TOLERANCE
			PERCENT ±
BLACK	0	---	---
BROWN	1	---	---
RED	2	---	*---
ORANGE	3	---	---
YELLOW	4	---	---
GREEN	5	---	---
BLUE	6	---	---
VIOLET	7	---	---
GRAY	8	---	---
WHITE	9	---	---
GOLD	---	0.1	5(J) *
SILVER	---	---	10(K) *
NO COLOR. . .	---	---	20(M) *

* SYMBOL DESIGNATION ALTERNATE FOR COLOR

20.373

SUMMARY OF RESISTOR TYPES

Type	Construction and size	Range of power ratings (watts)	Resistor range (ohms)	Other Characteristics	Applications	Military specifications
COMPOSITION						
Insulated	Cylindrical	1/10-2	3 – 10^7	100°C Max. temp. 5-201°C tolerance	General use	MIL-R-11A
Noninsulated	.5 × 2" .2 × 4" Diam.	1/10-2				
WIREWOUND						
Power-insulated	Cyl., flat, disk	2-200	.1 – 10^5	800°C Max. temp.	Motor control circuits	MIL-R-26B
Noninsulated	Cyl., 2 × 15"	125-1,200	.03-800	300°C Max. temp.	Standards, high precision acts	MIL-R-93A
Precision	Cyl., .5 × 1"	.1-4	.1-20 × 10^6			MIL-R-10683A
FILM						
Deposited carbon	Cyl., .3 × 1"	.1-2	10-5 × 10^6	100°C Max. temp.	Stable	
Metal	Cyl., .4 × 1"	.1-2	25-10^7	180°C Max. temp.	Stable at high temp.	MIL-R-10509A
High voltage	Cyl., 1 × 20"	0-10	1 Meg-10^7	200°C Max. temp.	Power supply divider	JAN-R-29

VARIABLE						
Composition	Toroidal element	0-2	0-4 Meg	120°C Max. temp.	General use	MIL-R-94A
Wirewound-precision	wiping arm	0-1	0-20,000	0-300°C	Instr., precise circuits	JAN-R-19
Wirewound-power		0-1,000	0-10,000	350°C Max.	Power control circuits	MIL-R-22A
SPECIAL						
Ballast tubes	Glass envelope	.5-5	—	Automatic	Regulators	
Vacuum type	Up to ST-19 bulb			Variable		
Semiconductors:	Solid-state	—	—	Specialized	Instrumentation	
Thermistors	transistor-type	—				
Varistor	mounting	—				
Binistor						

73

SUMMARY OF FIXED CAPACITOR TYPES

Dielectric classification	Capacity range (MF)	Voltage range	Leads, mounting	Operating temp. range (°C)	Application frequency and range
PAPER:					
Tubular	.0005–6.0	200–400	Axial	−55–+150	General purpose, up to 1 MHZ
Plastic film		200–600			
Special: bathtub, rectangular case	.1–2.0	100–12,500	Terminals		
MICA:					
Molded	.00001–.015	300–3000	Axial	−60–+85	General purpose, up to 10 MHZ
Potted plastic (ceramic case)	.0005–.1	250–5000	Terminals, ears, stud mounting		
Button	.00001–.0039	300	Tab and mounting stud		
CERAMIC:					
Tubular	.00001–.03	0 to 500 or 600	Radial or parallel wire	−55–+125	General purpose, 50–1000 MHZ
Disc					
Special: standoff feedthrough compensating	0–.00001 0–.00005		Body or stud mounting		UHF bypass, critically tuned **circuits, 50–1000 MHZ**
high voltage	.0005	10,000–20,000	Stud and terminal mounting		High-voltage TV filter 0–20 KHZ

ELECTROLYTIC:					
Tubular	0-100		Axial leads	0-+65	0-20 KHZ
Aluminum foil (plug-in)	6-2000				
Non-polarized (screw)		150-450	End mounting		
Wet or dry (base)	450-50	6-500			
Cylindrical (twist, prong)					
Tantalum:					
foil	.5-60	.5-150.0	Lead mounting	-55-+85	
tubular slug	7-300	.5-600.0	Body mounting	55-+200	
GLASS:					
Rectangular	.00005-.005	0-300	Axial	-60-+200	0-100 MHZ
VITREOUS ENAMEL:					
Rectangular	.00001-.01	0-300	Axial	-60-+200	0-100 MHZ
SPECIAL:					
Vacuum	.0001-.0005	0-50KV	Terminals	-60-+400	Blocking and decoupling
Gas	.00001-.00005	0-200	Terminals	-60-+200	Blocking and decoupling

CERAMIC CAPACITOR TYPES AND MARKINGS

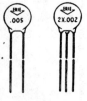

(a) Ceramic disc capacitors

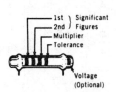

(b) Ceramic tubular capacitors

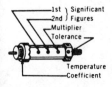

(c) Ceramic stand-off and feed-through capacitors

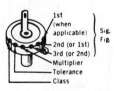

(d) Ceramic button capacitors

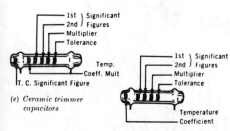

(e) Ceramic trimmer capacitors

(f) Ceramic trimmer capacitor using the piston principle

MOLDED MICA CAPACITOR
3 AND 6 DOT MARKINGS

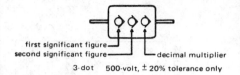

first significant figure
second significant figure
decimal multiplier

3-dot   500-volt, ± 20% tolerance only

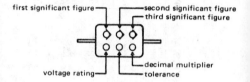

first significant figure
second significant figure
third significant figure
voltage rating
decimal multiplier
tolerance

MOLDED PAPER OR FOIL
CAPACITOR MARKINGS

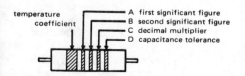

temperature coefficient

A first significant figure
B second significant figure
C decimal multiplier
D capacitance tolerance

DIODE SYMBOLS CHARACTERISTICS AND APPLICATIONS

Name	Symbol	Junction structure	Range Voltage–current frequency	Application usage
Diode Rectifier	Anode ▷⊢ Cathode	P N	$\mu A - 1000 \times A$ all frequencies	Rectification: – power – signal Demodulation
Zener (avalanche) diode		P N	0.5 → 50 watts 1 → 200 V DC and low frequencies	Regulation Voltage reference Clipping – limiting
Tunnel diode		P N	mW power Gc cutoff	Negative resistance High speed switching Oscillators Amplifiers
Varactor diode		P N	All frequencies mW power	Tuned resonant circuits Frequency multipliers Limiters

Conventional diodes

78

Special diodes	① Switching diodes ② Hot carrier diodes (Schottky) ③ Step recovery (Snap diodes) ④ P–I–N diodes ⑤ Mixer–detector diodes ⑥ Reed–oscillators diodes ⑦ Noise diodes ⑧ Avalanche diodes (Transit–time) ⑨ Backward diodes ⑩ 4 layer diodes (Thyristors)	⟶ (symbol)	P N With special: (a) Impurity treatment (b) Electrode configuration (c) Bias or voltage operation (d) Frequency	Mostly Low power and voltage for Communication and Data processing circuits	Computers ① ②④ Microwave receivers ⑤ ⑦ Instruments ⑥ ⑧ Optical measurements⑧ Modulators ⑨
	Light sensor diode	(symbol)	P N Transparent	Visible ⎫ Infrared ⎬ Light Ultraviolet ⎭	Transducers Illumination control Photo–switching
	Light emissive diodes (Injection lasers)	(symbol)	P N Transparent	Mostly infrared radiations	I. R. sensors Opto– ⎫ Amplification electronic ⎬ switching

TRANSISTOR AND THYRISTOR SYMBOLS AND CHARACTERISTICS

Name	Symbol	Junction structure	Range Voltage–current–frequency	Application usage
NPN transistor	Collector / Base / Emitter	Base / Collector / Emitter / N P N	Small 0–1 V / Large 1–10 Signal / Power to 100 W thru microwave frequency	Amplification Oscillation Power output Switching–detection
PNP transistor	Emitter / Base / Collector	Base / Collector / Emitter / P N P	Same as NPN	Same as NPN
Unijunction transistor (UJT)	Base 2 / Emitter / Base 1	Emitter / Base 2 / Base 1 / N P	Low power – under 1 watt Low voltage – less than 100 V LF: 0–300 KHz	Trigger / Timing / Oscillator / Comparator } Circuits
Field effect transistor (FET)	Drain / Gate / Source	Oxide / Source / Drain / Gate / P	Power to 50 watts Frequency to UHF High impedance G_m to 20,000 μ mhos	Amplifiers – oscillators Switches–gates Constant I generator Voltage controlled R.

Transistors

	Symbol	Structure	Ratings	Applications
Thyristors Silicon controlled rectifier (SCR)	Anode — Cathode / Gate	Gate C / P N P N / A	To 1000 A Low frequency DC – 50 KHz	Wide range of signal and power switching Motor phase control
Silicon controlled switch (SCS)	Anode gate — Cathode gate / Anode — Cathode	Gc C / P N P N / A GA	Low current switching circuits	Similar to SCR but with special restrictions stemming from two gate voltages
Gate turn-off switch (GTO)	Gate — Cathode / Anode	Gate C / P N P N / A	Similar to SCR with: Speed – 100 KHz Higher Power Voltage	Signal and power switching
Triac (bi-directional triode)	T_2 — Gate / T_1	T_2 / P N P N N_x / A Gate T_1	Current – 0 – 10 A High speed – 600 Hz	Bi-directional power switching and control Simple, flexible – otherwise like SCR
LASCR (light activated SCR)	Anode gate — Cathode / Anode	Gc C / P N P N / A	Current – 1 – 3 A Voltage – 100 – 200 V Low frequency as for SCR	Gate and light control switching Trigger for power SCR's Opto – electronic logic
LASCS (light activated switch)	Anode gate — Cathode / Anode — Cathode gate	Gc C / P N P N / A G	Power – 0.5 watts Low voltage – 0 – 50 V Low frequency as for SCR	Similar to LASCR

COAXIAL CABLE CHARACTERISTICS

(following basic specification MIL-C-17B)

Cable	Minimum dielectric strength (volts, rms)	Impedance (ohms)
RG–5B/U	7,000	50 ± 2
RG–6A/U	7,000	75 ± 3
RG–8A/U	10,000	50 ± 2
RG–9B/U	10,000	50 ± 2
RG–10A/U	10,000	50 ± 2
RG–11A/U	10,000	75 ± 3
RG–12A/U	10,000	75 ± 3
RG–13A/U	10,000	75 ± 3
RG–14A/U	12,000	50 ± 2
RG–17A/U	22,000	50 ± 2
RG–18A/U	11,000	50 ± 2
RG–19A/U	30,000	50 ± 2
RG–20A/U	30,000	50 ± 2
RG–21A/U	7,000	50 ± 2
RG–22B/U	2,000	95 ± 5
RG–23A/U	14,500	125 ± 5
RG–24A/U	14,500	125 ± 5
RG–25/U	12,000	48 ± 4
RG–25A/U	12,000	48 ± 4
RG–26/U	12,000	48 ± 4
RG–26A/U	12,000	48 ± 4
RG–27A/U	18,000	48 ± 4
RG–28/U	18,000	48 ± 4
RG–34A/U	15,000	75 ± 3
RG–35A/U	22,000	75 ± 3
RG–55A/U	5,000	50 ± 2
RG–57A/U	10,000	95 ± 5
RG–58C/U	5,000	50 ± 2
RG–59A/U	7,000	75 ± 3
RG–62A/U	3,000	93 ± 5
RG–63B/U	3,000	125 ± 6
RG–64/U	12,000	48 ± 4
RG–64A/U	12,000	48 ± 4
RG–65A/U	3,000	950 ± 50

Cable	Minimum dielectric strength (volts, rms)	Impedance (ohms)
RG–74A/U	12,000	50 ± 2
RG–79B/U	3,000	125 ± 6
RG–81/U	3,000	$50 + 4 \\ -2$
RG–82/U	5,000	$50 + 4 \\ -2$
RG–84A/U	22,000	75 ± 3
RG–85A/U	22,000	75 ± 3
RG–86/U	10,000	200 ± 10
RG–87A/U	10,000	50 ± 2
RG–88B/U	12,000	48 ± 4
RG–94A/U	12,000	50 ± 2
RG–108A/U	2,000	78 ± 7
RG–111A/U	2,000	95 ± 5
RG–114A/U	5,000	185 ± 10
RG–115/U	5,000	50 ± 2
RG–116/U	10,000	50 ± 2
RG–117/U	10,000	50 ± 2
RG–118/U	10,000	50 ± 2
RG–119/U	12,000	50 ± 2
RG–120/U	12,000	50 ± 2
RG–122/U	5,000	50 ± 2
RG–126/U	7,000	50 ± 2
RG–130/U	10,000	95 ± 5
RG–131/U	10,000	95 ± 5
RG–140/U	7,000	75 ± 3
RG–141/U	5,000	50 ± 2
RG–142/U	5,000	50 ± 2
RG–143/U	7,000	50 ± 2
RG–144/U	10,000	75 ± 3
RG–146/U	5,000	190 ± 10
RG–164/U	22,000	75 ± 3
RG–165/U	10,000	50 ± 2
RG–166/U	10,000	50 ± 2
RG–177/U	22,000	50 ± 2
RG–187/U	2,000	75 ± 3
RG–188/U	2,000	50 ± 2
RG–195/U	2,000	95 ± 3
RG–196/U	2,000	50 ± 2

TWO-CONDUCTOR SHIELDED CABLES

(following basic specification MIL-C-17B)

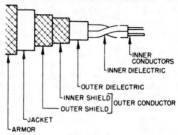

Figure. 1

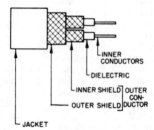

Figure. 2

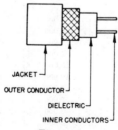

Figure. 3

TWO-CONDUCTOR SHIELDED CABLES
(Cont.)

RG-**/U no.	Fig.	Imp, ohms	Cap, μμf/ft	Dielectric strength, volts rms min.
22B	1	95 ± 5	16	2,000
23A	2	125 ± 5	12	14,500
24A	2	125 ± 5	12	14,500
57A	3	95 ± 5	17	10,000
86	3	200 ± 10	7.8	10,000
108A	1	78 ± 7	24.5	2,000
111A	1	95 ± 5	16	2,000
130	1	95 ± 5	—	10,000
131	1	95 ± 5	—	10,000

RG-**/U no.	Fig.	Inner conductor (each)			Dielectric	
					Inner	
		Strands /AWG	O.D. nom., inch	Mat. (1)	Mat. (2)	O.D. max.
22B	1	7/–	.040	C	A	.092
23A	2	7/–	.0855	C	A	.394
24A	2	7/–	.0855	C	A	.394
57A	3	7/–	.0855	C	–	—
86	3	7/–	.0855	C	–	—
108A	1	7/28	—	TC	A	.082
111A	1	7/–	.0456	C	A	.092
130	1	7/–	.0855	C	A	.193
131	1	7/–	.0855	C	A	.193

(Continued on next page)

** Insert number listed in column for desired cable.
1. Inner conductor material: C-Copper; TC-Tinned Copper.
2. Dielectric material: A-Polyethylene.
3. Jacket material: I-Black Synthetic Resin; IIa-Black noncontaminating synthetic resin.

TWO-CONDUCTOR SHIELDED CABLES
(Cont.)

RG-**/U no.	Fig.	Dielectric Outer		Outer conductor	
		Mat. (2)	O.D. max.	No. of shields	O.D. max., inch
22B	1	A	.291	2	.355
23A	2	–	–	3	–
24A	2	–	–	3	–
57A	3	A	.487	1	.540
86	3	A	–	–	–
108A	1	–	–	1	.177
111A	1	A	.291	2	.355
130	1	A	.487	1	.540
131	1	A	.487	1	.540

RG-**/U no.	Fig.	Protective coatings			O.D. max., inch
		Moisture seal	Jacket mat. (3)	Armor	
22B	1	—	II a	—	.430
23A	2	—	II a	—	.670 × .965 Oval
24A	2	—	II a	X	.735 × 1.034 Oval
57A	3	—	II a	—	.640
86	3	—	None	—	.665 × .315 Oval
108A	1	—	II a	—	.245
111A	1	—	II a	X	.490
130	1	—	I	—	.640
131	1	—	I	X	.710

** Insert number listed in column for desired cable.
1. Inner conductor material: C-Copper; TC-Tinned Copper.
2. Dielectric material: A-Polyethylene.
3. Jacket material: I-Black Synthetic Resin; IIa-Black noncontaminating synthetic resin.

RELAY TYPES AND DEFINITIONS

Types

Differential. A differential relay is a multiple-winding relay which operates when the current or voltage difference between the windings reaches a predetermined value.

High speed. A high speed relay is one which operates within 5 milliseconds.

Interlock. An interlock relay is a relay having two coils with their armatures and associated contacts so arranged that if one of the armatures is actuated, it prevents the other armature from being actuated until the first armature returns to its normal position.

Latch-in. A latch-in relay is one which is designed to lock the contacts in the energized position until the relay is either manually or electrically reset.

Marginal. A marginal relay is one which responds to make or break when the coil voltage or current reaches a predetermined value.

Polarized. A polarized relay is one which is responsive to the direction of current flow.

Ratchet. A ratchet relay is one which operates in cycles in accordance with a successive or predetermined arrangement of impulses.

Selector. A selector relay is one which permits the selection of one or more circuits from a number of circuits.

Sensitive. A sensitive relay is one which is designed to operate on 100 milliwatts or less.

Special purpose. A special purpose is a relay designed for a specific purpose or application that is not covered by any other type of relay defined herein.

Time delay. A time delay relay is one in which a delayed action is purposely introduced.

(Continued on next page)

Definitions

Operating frequency. The operating frequency is the rated frequency at which the relay coil is designed to operate.

Rated voltage. Rated voltage is the voltage at which the relay is designed to operate.

Rated coil current. Rated coil current is the current which flows through the coil when rated voltage is applied.

Pick-up voltage (or **current**). Pick-up voltage (or current) is the minimum value of voltage (or current) at which the relay is designed to operate.

SEMICONDUCTOR RECTIFIER TERMINOLOGY

Terminology	Symbol	Definition
Forward voltage drop	V_F	Instantaneous forward voltage drop during conduction of load current under stated conditions.
Full-cycle average forward voltage drop	$V_{Favg.}$	Forward voltage drop averaged over one complete cycle at stated temperature and with specified sinusoidal load current flowing in a 60-cps single-phase half-wave rectifier with resistive load.
Reverse (or leakage) current	i_R	Instantaneous value of reverse current at stated conditions of temperature and voltage.
Full-cycle average reverse current	$I_{Ravg.}$	Reverse current averaged over one complete cycle at specified PRV and temperature in a 60-cps sinusoidal single-phase half-wave rectifier with resistive load.
Thermal resistance	R_T	Steady-state temperature rise per unit power dissipation of the junction above the temperature of the external reference point. For stud mounted cells, thermal resistance is expressed in °C/Watt between junction and the bottom face of stud.
Transient thermal resistance	r_T	Apparent thermal resistance defining temperature rise per unit power dissipation of junction above reference

(Continued on next page)

SEMICONDUCTOR RECTIFIER
TERMINOLOGY (Cont.)

Terminology	Symbol	Definition
Transient thermal resistance (Cont.)	r_T	point for specified period of time after application of step function of heat to junction which was previously at equilibrium with ambient.
Cyclical peak reverse voltage or Repetitive peak inverse voltage	PRV or PIV	Maximum allowable instantaneous value of reverse voltage that may be applied across the rectifier repetitively under the specified conditions. Voltage may be applied as a sine wave, but it should not be construed to necessarily mean a continuous DC rating which is defined separately below. While this value of PRV does not represent a "breakdown" voltage, it should never be exceeded except by the transient rating if the device has such a rating.
Transient peak reverse voltage	PRV_{trans}	Maximum allowable instantaneous value of reverse voltage that may be applied on a nonrecurrent basis for the duration and conditions specified.
Continuous reverse DC (or blocking) voltage	V_{RDC}	Maximum voltage which the rectifier may block on a continuous basis. If this DC reverse voltage is applied immediately after a forward conducting cycle, other limitations may apply.

Terminology	Symbol	Definition
Maximum allowable DC output current	I_F	Maximum DC current which may be allowed to flow in the forward direction under stated conditions of temperature and reverse voltage. It is the average value of current through a cell when working in a single-phase circuit feeding a resistive or inductive load. A derating factor is usually applied for other conditions such as three-phase operation or capacitive loading.
Peak one-cycle surge (or fault) current	I_{surge}	Maximum allowable non-recurrent peak current of a single forward cycle (8.3 milliseconds duration) in a 60-cps single phase resistive load system. The surge may be preceded and followed by maximum rated voltage, current, and junction temperature conditions.
I-squared t	$I^2 t$	This is a measure of maximum forward nonrecurring overcurrent capability for very short pulses of specified duration. I is in rms amperes and t is pulse duration in seconds. The same conditions as listed above for I_{surge} apply.

GATE AND LOGIC SWITCHING SYMBOLS

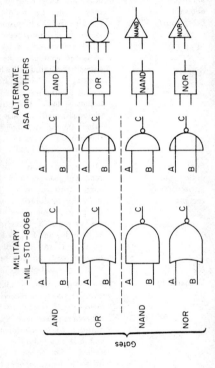

MILITARY -MIL-STD-806B

ALTERNATE ASA and OTHERS

Gates

AND

OR

NAND

NOR

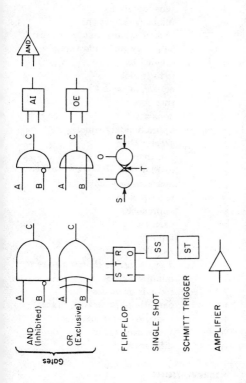

Gates

AND
(Inhibited)

OR
(Exclusive)

FLIP-FLOP

SINGLE SHOT   SS

SCHMITT TRIGGER   ST

AMPLIFIER

AND

AI

OE

93

MATHEMATICAL SYMBOLS AND ELECTRICAL ABBREVIATIONS

SYMBOLS

$+$	plus (addition)
$-$	minus (subtraction)
$\pm \; \mp$	plus or minus, (minus or plus)
$\times$	times, by (multiplication)
$\div, /$	divided by
$:$	is to (ratio)
$::$	equals, as, so is
$\therefore$	therefore
$=$	equals
$\sim \; \approx$	approximately equals
$>$	greater than
$<$	less than
$\geqq$	greater than or equals
$\leqq$	less than or equals
$\neq$	not equal to
$\doteq$	approaches
$\propto$	varies as
∞	infinity
$\parallel$	parallel to
$°$	degrees (arc or thermometer)
$'$	minutes or feet
$''$	seconds or inches
$(\;), [\;], \{\;\}$	parentheses, brackets, braces
$\angle, \perp$	angle, perpendicular to
Σ	summation of
Δ	difference

ABBREVIATIONS

AC	alternating current
a-f	audio-frequency
B	flux density
C	capacitance

cemf	instantaneous counter electromotive force
cm	centimeters
cps	cycles per second
C_T	total capacitance
d	distance between points
DC	direct-current
de	change in voltage
di	change in current
dq	change in charge
dt	change in time
E	voltage
e	instantaneous voltage
E_c	capacitive voltage
e_c	instantaneous capacitive voltage
E_L	inductive voltage
e_L	instantaneous inductive voltage
E_m	maximum voltage
E_{max}	maximum voltage
emf	electromotive force
E_p	primary voltage
E_s	secondary voltage
f	frequency
f_r	frequency at resonance
H	magnetizing flux
h	henry
I	current
i	instantaneous current
I_c	capacitive current
i_c	instantaneous capacitive current
I_{eff}	effective current
IF	intermediate-frequency
I_L	inductive current
i_L	instantaneous inductive current
I_{min}	minimum current

(Continued on next page)

I_{max}	maximum current
I_p	plate current
I_R	current through resistance
i_R	instaneous current through resistance
I_s	secondary current
I_T	total current
$I\phi$	phase current
K	coefficient of coupling
kc	kilocycle
L	inductance
L-C	inductance-capacitance
L-C-R	inductance-capacitance-resistance
L_T	total inductance
mh	millihenry
N	revolutions per minute
N_p	primary turns
N_s	secondary turns
P	power
p	instantaneous power
P_{ap}	apparent power
P_{av}	average power
P_p	primary power
P_s	secondary power
Q	charge or quality
q	instantaneous charge
RF	radio-frequency
R_G	grid resistance
R_0	load resistance
rpm	revolutions per minute
sq cm	square centimeters
t	time constant
t	time (seconds)

μf	microfarad
$\mu\mu$f	micromicrofarad
V, v	volt
X_c	capacitive reactance
X_L	inductive reactance
Z	impedance
Z_0	load impedance
Z_p	primary impedance
Z_s	secondary impedance
Z_T	total impedance

GREEK ALPHABET
AND
COMMON PREFIXES

Greek Alphabet

Name	Capital	Lower case	Designates
Alpha	A	α	Angles
Beta	B	β	Angles, flux density
Gamma	Γ	γ	Conductivity
Delta	Δ	δ	Variation of a quantity, increment
Epsilon	E	ϵ	Base of natural logarithms (2.71828)
Zeta	Z	ζ	Impedance, coefficients, coordinates
Eta	H	η	Hysteresis coefficient, efficiency
Theta	Θ	θ	Phase angle
Iota	I	ι	
Kappa	K	κ	Dielectric constant, coupling coefficient, susceptibility
Lambda	Λ	λ	Wavelength
Mu	M	μ	Permeability, micro, amplification factor
Nu	N	ν	Reluctivity
Xi	Ξ	ξ	
Omicron	O	o	
Pi	Π	π	3.1416
Rho	P	ρ	Resistivity
Sigma	Σ	σ	
Tau	T	τ	Time constant, time-phase displacement
Upsilon	Υ	υ	
Phi	Φ	ϕ	Angles, magnetic flux
Chi	X	χ	
Psi	Ψ	ψ	Dielectric flux, phase difference
Omega	Ω	ω	Ohms (capital), angular velocity ($2\pi f$)

Common Prefixes

Metric prefix	Meaning		Associated with
Mega	Million	(1,000,000)	Volt, ohms, cycles, amperes
Kilo	Thousand	(1,000)	Volts, watts, cycles, meters, amperes
Hecto	Hundred	(100)	Meters
Deka	Ten	(10)	Meters
Deci	One-tenth	(0.1)	Meters
Centi	One-hundredth	(0.01)	Meters
Milli	One-thousandth	(0.001)	Volts, amperes, meters, henrys, watts, ohms
Micro	One-millionth	(0.0000001)	Volts, amperes, farads, henrys, mhos, ohms
Micromicro	One-millionth of one-millionth	(0.000,000,000,001)	Volts, amperes, farads, coulombs

ELECTRICAL AND PHYSICAL SYMBOLS

Quantity	Symbol	Equation
Length	l	
Distance	d	
Mass	m	
Time	t	
Velocity	v	$v = l/t$
Acceleration	α	$\alpha = v/t$
Force	F	$F = \eta\alpha$
Work	W	$W = Fl$
Power	P	$P = w/t$
Permittivity of medium	ϵ	
Charge	q	$F = q_1 q_2/E\,r^2$
Capacitance	C	$C = q/v$
Potential difference	V or E	$V = \dfrac{W}{q}$
emf	e	$e = -d\phi/dt$
Current	I	$I = dq/dt$
Resistance	R	$R = V/I$
Resistivity	ρ	
Conductance	G	$G = 1/R$
Conductivity	γ	$\gamma = 1/\mu$
Permeability	μ	
Reluctivity	ν	$\nu = 1/\mu$
Pole strength	m	$F = m_1 m_2/\mu\tau^2$
Magnetomotive force	F	
Magnetizing force	H	$H = F/I$
Magnetic flux density	B	$B = \mu H$
Magnetic flux	ϕ	$\phi = BA$
Reluctance	R	$R = F/\phi$
Inductance	L	$L = e/(dI/dt)$

Practical unit	Subrationalized mks.
centimeter	Meter
centimeter	Meter
	Kilogram
second	Second
cm/sec	Meter/sec
cm/sec^2	Meter/sec^2
	$\dfrac{\text{joule}}{\text{meter}} = \text{newton}$
joule	Joule
watt	Watt
$\dfrac{1}{(9 \times 10^N)}$ farad/cm	$\dfrac{1}{(36\pi \times 10^9)}$ farad/meter
coulomb	Coulomb
farad	Farad
volt	Volt
volt	Volt
ampere	Ampere
ohm	ohm
ohm/cm	Ohm/meter
mho	mho
mho/cm	mho/meter
10^{-9} henry/cm	$\dfrac{4\pi \times 10^{-7} \text{ henry}}{\text{meter}}$
	Weber
$\frac{1}{4}\pi$ ampere turn	Ampere turn
$\frac{1}{4}\pi$ ampere turn	Ampere turn/m
weber/cm^2	Weber/meter^2
weber or volt-sec	Weber = volt-sec
$\dfrac{\frac{1}{4}\pi \text{ ampere turn}}{\text{weber}}$	$\dfrac{\text{ampere turn}}{\text{weber}}$
henry	Henry

DECIMAL–BINARY NUMBER CONVERSION

Decimal	Binary	Decimal	Binary
0	0 0 0 0	1 2	1 1 0 0
1	0 0 0 1	1 3	1 1 0 1
2	0 0 1 0	1 4	1 1 1 0
3	0 0 1 1	1 5	1 1 1 1
4	0 1 0 0	1 6	1 0 0 0 0
5	0 1 0 1	1 7	1 0 0 0 1
6	0 1 1 0	1 8	1 0 0 1 0
7	0 1 1 1	1 9	1 0 0 1 1
8	1 0 0 0	2 0	1 0 1 0 0
9	1 0 0 1	1 0 0	1 1 0 0 1 0 0
1 0	1 0 1 0	1 0 2 3	1 1 1 1 1 1 1 1 1 1
1 1	1 0 1 1		

COMPUTER NUMBER CODES

Decimal	BCD	7421 Code	Gray Code	X5-3 Code	2-out-of-5 Code
0	0000	0000	0000	0011	00110
1	0001	0001	0001	0100	00011
2	0010	0010	0011	0101	00101
3	0011	0011	0010	0110	01001
4	0100	0100	0110	0111	01010
5	0101	0101	0111	1000	01100
6	0110	0110	0101	1001	10001
7	0111	1000	0100	1010	10010
8	1000	1001	1100	1011	10100
9	1001	1010	1101	1100	11000

NUMBER FUNCTIONS

No.	Square	Cube	Square root	Cube root	No. = Diam.		No.
					Circum.	Area	
1	1	1	1.0000	1.0000	3.142	0.7854	1
2	4	8	1.4142	1.2599	6.283	3.1416	2
3	9	27	1.7321	1.4423	9.425	7.0686	3
4	16	64	2.0000	1.5874	12.566	12.5664	4
5	25	125	2.2361	1.7100	15.708	19.6350	5
6	36	216	2.4495	1.8171	18.850	28.2743	6
7	49	343	2.6458	1.9129	21.991	38.4845	7
8	64	512	2.8284	2.0000	25.133	50.2655	8
9	81	729	3.0000	2.0801	28.274	63.6173	9
10	100	1000	3.1623	2.1544	31.416	78.5398	10
11	121	1331	3.3166	2.2240	34.558	95.0332	11
12	144	1728	3.4641	2.2894	37.699	113.097	12
13	169	2197	3.6056	2.3513	40.841	132.732	13
14	196	2744	3.7417	2.4101	43.982	153.938	14

(Continued on next page)

NUMBER FUNCTION (Cont.)

No.	Square	Cube	Square root	Cube root	No. = Diam. Circum.	No. = Diam. Area	No.
15	225	3375	3.8730	2.4662	47.124	176.715	15
16	256	4096	4.0000	2.5198	50.265	201.062	16
17	289	4913	4.1231	2.5713	53.407	226.980	17
18	324	5832	4.2426	2.6207	56.549	254.469	18
19	361	6859	4.3589	2.6684	59.690	283.529	19
20	400	8000	4.4721	2.7144	62.832	314.159	20
21	441	9261	4.5826	2.7589	65.973	346.361	21
22	484	10648	4.6904	2.8020	69.115	380.133	22
23	529	12167	4.7958	2.8439	72.257	415.476	23
24	576	13824	4.8990	2.8845	75.398	452.389	24
25	625	15625	5.0000	2.9240	78.540	490.874	25
26	676	17576	5.0990	2.9625	81.681	530.929	26
27	729	19683	5.1962	3.0000	84.823	572.555	27
28	784	21952	5.2915	3.0366	87.965	615.752	28
29	841	24389	5.3852	3.0723	91.106	660.520	29

n	n²	n³	√n	∛n			n
30	900	27000	5.4772	3.1072	94.248	706.858	30
31	961	29791	5.5678	3.1414	97.389	754.768	31
32	1024	32768	5.6569	3.1748	100.531	804.248	32
33	1089	35937	5.7446	3.2075	103.673	855.299	33
34	1156	39304	5.8310	3.2396	106.814	907.920	34
35	1225	42875	5.9161	3.2711	109.956	962.113	35
36	1296	46656	6.0000	3.3019	113.097	1017.88	36
37	1369	50653	6.0828	3.3322	116.239	1075.21	37
38	1444	54872	6.1644	3.3620	119.381	1134.11	38
39	1521	59319	6.2450	3.3912	122.522	1194.59	39
40	1600	64000	6.3246	3.4200	125.66	1256.64	40
41	1681	68921	6.4031	3.4482	128.81	1320.25	41
42	1764	74088	6.4807	3.4760	131.95	1385.44	42
43	1849	79507	6.5574	3.5034	135.09	1452.20	43
44	1936	85184	6.6332	3.5303	138.23	1520.53	44
45	2025	91125	6.7082	3.5569	141.37	1590.43	45
46	2116	97336	6.7823	3.5830	144.51	1661.90	46
47	2209	103823	6.8557	3.6088	147.65	1734.94	47
48	2304	110592	6.9282	3.6342	150.80	1809.56	48

(Continued on next page)

NUMBER FUNCTION (Cont.)

No.	Square	Cube	Square root	Cube root	Circum.	Area	No.
					No. = Diam.		
49	2401	117649	7.0000	3.6593	153.94	1885.74	49
50	2500	125000	7.0711	3.6840	157.08	1963.50	50
51	2601	132651	7.1414	3.7084	160.22	2042.82	51
52	2704	140608	7.2111	3.8325	163.36	2123.72	52
53	2809	148877	7.2801	3.7563	166.50	2206.18	53
54	2916	157464	7.3485	3.7798	169.65	2290.22	54
55	3025	166375	7.4162	3.8030	172.79	2375.83	55
56	3136	175616	7.4833	3.8259	175.93	2463.01	56
57	3249	185193	7.5498	3.8485	179.07	2551.76	57
58	3364	195112	7.6158	3.8709	182.21	2642.08	58
59	3481	205379	7.6811	3.8930	185.35	2733.97	59
60	3600	216000	7.7460	3.9149	188.50	2827.43	60
61	3721	226981	7.8102	3.9365	191.64	2922.47	61
62	3844	238328	7.8740	3.9579	194.78	3019.07	62
63	3969	250047	7.9373	3.9791	197.92	3117.25	63
64	4096	262114	8.0000	4.0000	201.06	3216.99	64

n	n²	n³					n
65	4225	274625	8.0623	4.0207	204.20	3318.31	65
66	4356	287496	8.1240	4.0412	207.35	3421.19	66
67	4489	300763	8.1854	4.0615	210.49	3525.65	67
68	4624	314432	8.2462	4.0817	213.63	3631.68	68
69	4761	328509	8.3066	4.1016	216.77	3739.28	69
70	4900	343000	8.3666	4.1213	219.91	3848.45	70
71	5041	357911	8.4261	4.1408	223.05	3959.19	71
72	5184	373248	8.4853	4.1602	226.19	4071.50	72
73	5329	389017	8.5440	4.1793	229.34	4185.39	73
74	5476	405224	8.6023	4.1983	232.48	4300.84	74
75	5625	421875	8.6603	4.2172	235.62	4417.86	75
76	5776	438976	8.7178	4.2358	238.76	4536.46	76
77	5929	456533	8.7750	4.2543	241.90	4656.63	77
78	6084	474552	8.8318	4.2727	245.04	4778.36	78
79	6241	493039	8.8882	4.2908	248.19	4901.67	79
80	6400	512000	8.9443	4.3089	251.33	5026.55	80
81	6561	531441	9.0000	4.3267	254.47	5153.00	81
82	6724	551368	9.0554	4.3445	257.61	5281.02	82
83	6889	571787	9.1104	4.3621	260.75	5410.61	83

(Continued on next page)

NUMBER FUNCTION (Cont.)

No.	Square	Cube	Square root	Cube root	No. = Diam.		No.
					Circum.	Area	
84	7056	592704	9.1652	4.3795	263.89	5541.77	84
85	7225	614125	9.2195	4.3968	267.04	5674.50	85
86	7396	636056	9.2736	4.4140	270.18	5808.80	86
87	7569	658503	9.3274	4.4310	273.32	5944.68	87
88	7744	681472	9.3808	4.4480	276.46	6082.12	88
89	7921	704969	9.4340	4.4647	279.60	6221.14	89
90	8100	729000	9.4868	4.4814	282.74	6361.73	90
91	8281	753571	9.5394	4.4979	285.88	6503.88	91
92	8464	778688	9.5917	4.5144	289.03	6647.61	92
93	8649	804357	9.6437	4.5307	292.17	6792.91	93
94	8836	830584	9.6954	4.5468	295.31	6939.78	94
95	9025	857375	9.7468	4.5629	298.45	7088.22	95
96	9216	884736	9.7980	4.5789	301.59	7238.23	96
97	9409	912673	9.8489	4.5947	304.73	7389.81	97
98	9604	941192	9.8995	4.6104	307.88	7542.96	98
99	9801	970299	9.9499	4.6261	311.02	7697.69	99

FREQUENTLY USED ANGLES AND THEIR FUNCTIONS

Angle	$\sin A$	$\cos A$	$\tan A$	$\cot A$	$\sec A$	$\csc A$
0°	0	1	0	∞	1	∞
30°	$\dfrac{1}{2}$	$\dfrac{\sqrt{3}}{2}$	$\dfrac{\sqrt{3}}{3}$	$\sqrt{3}$	$\dfrac{2\sqrt{3}}{3}$	2
45°	$\dfrac{\sqrt{2}}{2}$	$\dfrac{\sqrt{2}}{2}$	1	1	$\sqrt{2}$	$\sqrt{2}$
60°	$\dfrac{\sqrt{3}}{2}$	$\dfrac{1}{2}$	$\sqrt{3}$	$\dfrac{\sqrt{3}}{3}$	2	$\dfrac{2\sqrt{3}}{3}$
90°	1	0	∞	0	∞	1
120°	$\dfrac{\sqrt{3}}{2}$	$-\dfrac{1}{2}$	$-\sqrt{3}$	$-\dfrac{\sqrt{3}}{3}$	-2	$\dfrac{2\sqrt{3}}{3}$
180°	0	-1	0	∞	-1	∞
270°	-1	0	∞	0	∞	-1
360°	0	1	0	∞	1	∞

ANGLE FUNCTION CONVERSION

Function	sine	cosine	tangent	cotangent	secant	cosecant
Sin A	$\sin A$	$\pm\sqrt{1-\cos^2 A}$	$\dfrac{\tan A}{\pm\sqrt{1+\tan^2 A}}$	$\dfrac{1}{\pm\sqrt{1+\cot^2 A}}$	$\dfrac{\pm\sqrt{\sec^2 A-1}}{\sec A}$	$\dfrac{1}{\csc A}$
Cos A	$\pm\sqrt{1-\sin^2 A}$	$\cos A$	$\dfrac{1}{\pm\sqrt{1+\tan^2 A}}$	$\dfrac{\cot A}{\pm\sqrt{1+\cot^2 A}}$	$\dfrac{1}{\sec A}$	$\dfrac{\pm\sqrt{\csc^2 A-1}}{\csc A}$
Tan A	$\dfrac{\sin A}{\pm\sqrt{1-\sin^2 A}}$	$\dfrac{\pm\sqrt{1-\cos^2 A}}{\cos A}$	$\tan A$	$\dfrac{1}{\cot A}$	$\pm\sqrt{\sec^2 A-1}$	$\dfrac{1}{\pm\sqrt{\csc^2 A-1}}$
Cot A	$\dfrac{\pm\sqrt{1-\sin^2 A}}{\sin A}$	$\dfrac{\cos A}{\pm\sqrt{1-\cos^2 A}}$	$\dfrac{1}{\tan A}$	$\cot A$	$\dfrac{1}{\pm\sqrt{\sec^2 A-1}}$	$\pm\sqrt{\csc^2 A-1}$
Sec A	$\dfrac{1}{\pm\sqrt{1-\sin^2 A}}$	$\dfrac{1}{\cos A}$	$\pm\sqrt{1+\tan^2 A}$	$\dfrac{\pm\sqrt{1+\cot^2 A}}{\cot A}$	$\sec A$	$\dfrac{\csc A}{\pm\sqrt{\csc^2 A-1}}$
Csc A	$\dfrac{1}{\sin A}$	$\dfrac{1}{\pm\sqrt{1-\cos^2 A}}$	$\dfrac{\pm\sqrt{1+\tan^2 A}}{\tan A}$	$\pm\sqrt{1+\cot^2 A}$	$\dfrac{\sec A}{\pm\sqrt{\sec^2 A-1}}$	$\csc A$

POWERS OF NUMBERS

n	n^4	n^5	n^6	n^7	n^8
1	1	1	1	1	1
2	16	32	64	128	256
3	81	243	729	2187	6561
4	256	1024	4096	16384	65536
5	625	3125	15625	78125	390625
6	1296	7776	46656	279936	1679616
7	2401	16807	117649	823543	5764801
8	4096	32768	262144	2097152	16777216
9	6561	59049	531441	4782969	43046721
					$\times 10^8$
10	10000	100000	1000000	10000000	1.000000
11	14641	161051	1771561	19487171	2.143589
12	20736	248832	2985984	35831808	4.299817
13	28561	371293	4826809	62748517	8.157307
14	38416	537824	7529536	105413504	14.757891
15	50625	759375	11390625	170859375	25.628906
16	65536	1048576	16777216	268435456	42.949673
17	83521	1419857	24137569	410338673	69.757574
18	104976	1889568	34012224	612220032	110.199606
19	130321	2476099	47045881	893871739	169.835630
				$\times 10^9$	$\times 10^{10}$
20	160000	3200000	64000000	1.280000	2.560000
21	194481	4084101	85766121	1.801089	3.782286
22	234256	5153632	113379904	2.494358	5.487587
23	279841	6436343	148035889	3.404825	7.831099
24	331776	7962624	191102976	4.586471	11.007531
25	390625	9765625	244140625	6.103516	15.258789
26	456976	11881376	308915776	8.031810	20.882706
27	531441	14348907	387420489	10.460353	28.242954
28	614656	17210368	481890304	13.492929	37.780200
29	707281	20511149	594823321	17.249876	50.024641
		$\times 10^8$		$\times 10^{10}$	$\times 10^{11}$
30	810000	24300000	7.290000	2.187000	6.561000
31	923521	28629151	8.875037	2.751261	8.528910
32	1048576	33554432	10.737418	3.435974	10.995116
33	1185921	39135393	12.914680	4.261844	14.064086
34	1336336	45435424	15.448044	5.252335	17.857939
35	1500625	52521875	18.382656	6.433930	22.518754
36	1679616	60466176	21.767823	7.836416	28.211099
37	1874161	69343957	25.657264	9.493188	35.124795
38	2085136	79235168	30.109364	11.441558	43.477921
39	2313441	90224199	35.187438	13.723101	53.520093

(Continued on next page)

111

POWERS OF NUMBERS (Cont.)

n	n^4	n^5	n^6	n^7	n^8
			$\times 10^9$	$\times 10^{10}$	$\times 10^{12}$
40	2560000	102400000	4.096000	16.384000	6.553600
41	2825761	115856201	4.750104	19.475427	7.984925
42	3111696	130691232	5.489032	23.053933	9.682652
43	3418801	147008443	6.321363	27.181861	11.688200
44	3748096	164916224	7.256314	31.927781	14.048224
45	4100625	184528125	8.303766	37.366945	16.815125
46	4477456	205962976	9.474297	43.581766	20.047612
47	4879681	229345007	10.779215	50.662312	23.811287
48	5308416	254803968	12.230590	58.706834	28.179280
49	5764801	282475249	13.841287	67.822307	33.232931
50	6250000	312500000	15.625000	78.125000	39.062500
			$\times 10^9$	$\times 10^{11}$	$\times 10^{13}$
50	6250000	312500000	15.625000	7.812500	3.906250
51	6765201	345025251	17.596288	8.974107	4.576794
52	7311616	380204032	19.770616	10.280717	5.345973
53	7890481	418195493	22.164361	11.747111	6.225969
54	8503056	459165024	24.794911	13.389252	7.230196
55	9150625	503284375	27.680641	15.224352	8.373394
56	9834496	550731776	30.840979	17.270948	9.671731
57	10556001	601692057	34.296447	19.548975	11.142916
58	11316496	656356768	38.068693	22.079842	12.806308
59	12117361	714924299	42.180534	24.886515	14.683044
		$\times 10^8$	$\times 10^{10}$	$\times 10^{11}$	$\times 10^{13}$
60	12960000	7.776000	4.665600	27.993600	16.796160
61	13845841	8.445963	5.152037	31.427428	19.170731
62	14776336	9.161328	5.680024	35.216146	21.834011
63	15752961	9.924365	6.252350	39.389806	24.815578
64	16777216	10.737418	6.871948	43.980465	28.147498
65	17850625	11.602906	7.541889	49.022279	31.864481
66	18974736	12.523326	8.265395	54.551607	36.004061
67	20151121	13.501251	9.045838	60.607116	40.606768
68	21381376	14.539336	9.886748	67.229888	45.716324
69	22667121	15.640313	10.791816	74.463533	51.379837
		$\times 10^8$	$\times 10^{10}$	$\times 10^{12}$	$\times 10^{14}$
70	24010000	16.807000	11.764900	8.235430	5.764801
71	25411681	18.042294	12.810028	9.095120	6.457535
72	26873856	19.349176	13.931407	10.030613	7.222041
73	28398241	20.730716	15.133423	11.047399	8.064601
74	29986576	22.190066	16.420649	12.151280	8.991947
75	31640625	23.730469	17.797852	13.348389	10.011292
76	33362176	25.355254	19.269993	14.645195	11.130348
77	35153041	27.067842	20.842238	16.048523	12.357363
78	37015056	28.871744	22.519960	17.565569	13.701144
79	38950081	30.770564	24.308746	19.203909	15.171088

n	n^4	n^5	n^6	n^7	n^8
		$\times 10^8$	$\times 10^{10}$	$\times 10^{12}$	$\times 10^{14}$
80	40960000	32.768000	26.214400	20.971520	16.777216
81	43046721	34.867844	28.242954	22.876792	18.530202
82	45212176	37.073984	30.400667	24.928547	20.441409
83	47458321	39.390406	32.694037	27.136051	22.522922
84	49787136	41.821194	35.129803	29.509035	24.787589
85	52200625	44.370531	37.714952	32.057709	27.249053
86	54700816	47.042702	40.456724	34.792782	29.921793
87	57289761	49.842092	43.362620	37.725479	32.821167
88	59969536	52.773192	46.440409	40.867560	35.963452
89	62742241	55.840594	49.698129	44.231335	39.365888
		$\times 10^9$	$\times 10^{11}$	$\times 10^{13}$	$\times 10^{15}$
90	65610000	5.904900	5.314410	4.782969	4.304672
91	68574961	6.240321	5.678693	5.167610	4.702525
92	71639296	6.590815	6.063550	5.578466	5.132189
93	74805201	6.956884	6.469902	6.017009	5.595818
94	78074896	7.339040	6.898698	6.484776	6.095689
95	81450625	7.737809	7.350919	6.983373	6.634204
96	84934656	8.153727	7.827578	7.514475	7.213896
97	88529281	8.587340	8.329720	8.079828	7.837434
98	92236816	9.039208	8.858424	8.681255	8.507630
99	96059601	9.509900	9.414801	9.320653	9.227447
100	100000000	10.000000	10.000000	10.000000	10.000000

TABLE OF SINES, COSINES AND TANGENTS

Angle	Radians	Sine	Cosine	Tangent
0°	.0000	.0000	1.0000	.0000
1	.0175	.0175	.9998	.0175
2	.0349	.0349	.9994	.0349
3	.0524	.0523	.9986	.0524
4	.0698	.0698	.9976	.0699
5	.0873	.0872	.9962	.0875
6	.1047	.1045	.9945	.1051
7	.1222	.1219	.9925	.1228
8	.1396	.1392	.9903	.1405
9	.1571	.1564	.9877	.1584
10	.1745	.1736	.9848	.1763
11	.1920	.1908	.9816	.1944
12	.2094	.2079	.9781	.2126
13	.2269	.2250	.9744	.2309
14	.2443	.2419	.9703	.2493
15	.2618	.2588	.9659	.2679
16	.2793	.2756	.9613	.2867
17	.2967	.2924	.9563	.3057
18	.3142	.3090	.9511	.3249
19	.3316	.3256	.9455	.3443
20	.3491	.3420	.9397	.3640
21	.3665	.3584	.9336	.3839
22	.3840	.3746	.9272	.4040
23	.4014	.3907	.9205	.4245
24	.4189	.4067	.9135	.4452
25	.4363	.4226	.9063	.4663
26	.4538	.4384	.8988	.4877
27	.4712	.4540	.8910	.5095
28	.4887	.4695	.8829	.5317
29	.5061	.4848	.8746	.5543
30	.5236	.5000	.8660	.5774
31	.5411	.5150	.8572	.6009
32	.5585	.5299	.8480	.6249
33	.5760	.5446	.8387	.6494
34	.5934	.5592	.8290	.6745
35	.6109	.5736	.8192	.7002

Angle	Radians	Sine	Cosine	Tangent
36	.6283	.5878	.8090	.7265
37	.6458	.6018	.7986	.7536
38	.6632	.6157	.7880	.7813
39	.6807	.6293	.7771	.8098
40	.6981	.6428	.7660	.8391
41	.7156	.6561	.7547	.8693
42	.7330	.6691	.7431	.9004
43	.7505	.6820	.7314	.9325
44	.7679	.6947	.7193	.9657
45	.7854	.7071	.7071	1.0000
46	.8029	.7193	.6947	1.0355
47	.8203	.7314	.6820	1.0724
48	.8378	.7431	.6691	1.1106
49	.8552	.7547	.6561	1.1504
50	.8727	.7660	.6428	1.1918
51	.8901	.7771	.6293	1.2349
52	.9076	.7880	.6157	1.2799
53	.9250	.7986	.6018	1.3270
54	.9425	.8090	.5878	1.3764
55	.9599	.8192	.5736	1.4281
56	.9774	.8290	.5592	1.4826
57	.9948	.8387	.5446	1.5399
58	1.0123	.8480	.5299	1.6003
59	1.0297	.8572	.5150	1.6643
60	1.0472	.8660	.5000	1.7321
61	1.0647	.8746	.4848	1.8040
62	1.0821	.8829	.4695	1.8807
63	1.0996	.8910	.4540	1.9626
64	1.1170	.8988	.4384	2.0503
65	1.1345	.9063	.4226	2.1445
66	1.1519	.9135	.4067	2.2460
67	1.1694	.9205	.3907	2.3559
68	1.1868	.9272	.3746	2.4751
69	1.2043	.9336	.3584	2.6051
70	1.2217	.9397	.3420	2.7475

(Continued on next page)

TABLE OF SINES, COSINES AND TANGENTS (Cont.)

Angle	Radians	Sine	Cosine	Tangent
71	1.2392	.9455	.3256	2.9042
72	1.2566	.9511	.3090	3.0777
73	1.2741	.9563	.2924	3.2709
74	1.2915	.9613	.2756	3.4874
75	1.3090	.9659	.2588	3.7321
76	1.3265	.9703	.2419	4.0108
77	1.3439	.9744	.2250	4.3315
78	1.3614	.9781	.2079	4.7046
79	1.3788	.9816	.1908	5.1446
80	1.3963	.9848	.1736	5.6713
81	1.4137	.9877	.1564	6.3138
82	1.4312	.9903	.1392	7.1154
83	1.4486	.9925	.1219	8.1443
84	1.4661	.9945	.1045	9.5144
85	1.4835	.9962	.0872	11.43
86	1.5010	.9976	.0698	14.30
87	1.5184	.9986	.0523	19.08
88	1.5359	.9994	.0349	28.64
89	1.5533	.9998	.0175	57.29

FRACTIONS AND DECIMAL EQUIVALENTS

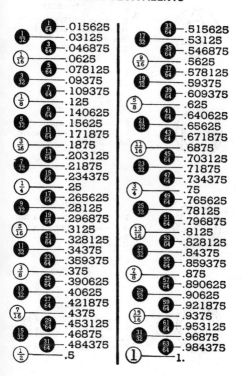

Fraction	Decimal
1/64	.015625
1/32	.03125
3/64	.046875
1/16	.0625
5/64	.078125
3/32	.09375
7/64	.109375
1/8	.125
9/64	.140625
5/32	.15625
11/64	.171875
3/16	.1875
13/64	.203125
7/32	.21875
15/64	.234375
1/4	.25
17/64	.265625
9/32	.28125
19/64	.296875
5/16	.3125
21/64	.328125
11/32	.34375
23/64	.359375
3/8	.375
25/64	.390625
13/32	.40625
27/64	.421875
7/16	.4375
29/64	.453125
15/32	.46875
31/64	.484375
1/2	.5
33/64	.515625
17/32	.53125
35/64	.546875
9/16	.5625
37/64	.578125
19/32	.59375
39/64	.609375
5/8	.625
41/64	.640625
21/32	.65625
43/64	.671875
11/16	.6875
45/64	.703125
23/32	.71875
47/64	.734375
3/4	.75
49/64	.765625
25/32	.78125
51/64	.796875
13/16	.8125
53/64	.828125
27/32	.84375
55/64	.859375
7/8	.875
57/64	.890625
29/32	.90625
59/64	.921875
15/16	.9375
61/64	.953125
31/32	.96875
63/64	.984375
1	1.

COMMON GEOMETRIC RELATIONSHIPS

TRIANGLE

E = side x .57735
D = side x 1.1547 = 2 E
Side = D x .866
C = E x .5 = D x .25

SQUARE

E = side = D x .7071
D = side x 1.4142 = diagonal
Side = D x .7071
C = D x .14645

PENTAGON

E = side x 1.3764 = D x .809
D = side x 1.7013 = E x 1.2361
Side = D x .5878
C = D x .0955

HEXAGON

E = side x 1.7321 = D x .866
D = side x 2 = E x 1.1547
Side = D x .5
C = D x .067

OCTAGON

E = side x 2.4142 = D x .9239
D = side x 2.6131 = E x 1.0824
Side = D x .3827
C = D x .038

TRIGONOMETRIC SOLUTION OF TRIANGLES

TO FIND	GIVEN	FORMULAE
A	BC	180°-(B+C)
TAN A	ABC	$\frac{A \times SIN\ C}{B-A \times COS\ C}$
COS A	ABC	$\frac{B^2+C^2-A^2}{2BC}$
SIN A	ACC	$\frac{A \cdot SIN\ C}{C}$
SIN A	ABB	$\frac{A \cdot SIN\ B}{B}$
TAN A	ACB	$\frac{A \cdot SIN\ B}{C-(A \cdot COS\ B)}$
B	AC	180°-(A+C)
SIN B	ABA	$\frac{B \cdot SIN\ A}{A}$
COS B	ABC	$\frac{C^2+A^2-B^2}{2AC}$
TAN B	BCA	$\frac{B \cdot SIN\ A}{C-(B \cdot COS\ A)}$
SIN B	BCC	$\frac{B \cdot SIN\ C}{C}$

TO FIND	GIVEN	FORMULAE
A	CAC	$\frac{C \cdot SIN\ A}{SIN\ C}$
A	BAB	$\frac{B \cdot SIN\ A}{SIN\ B}$
A	BCA	$\sqrt{B^2+C^2-(2BC \cdot COS\ A)}$
B	AAB	$\frac{A \cdot SIN\ B}{SIN\ A}$
B	CBC	$\frac{C \cdot SIN\ B}{SIN\ C}$
B	ACB	$\sqrt{C^2+A^2-(2AC \cdot COS\ B)}$

TO FIND	GIVEN	FORMULAE
C	AB	180°-(A+B)
SIN C	ACA	$\frac{C \cdot SIN\ A}{A}$
TAN C	BCA	$\frac{C \cdot SIN\ A}{B-(C \cdot COS\ A)}$
SIN C	BCB	$\frac{C \cdot SIN\ B}{B}$
TAN C	ABC	$\frac{C \cdot SIN\ B}{A-(C \cdot COS\ B)}$
COS C	ABC	$\frac{A^2+B^2-C^2}{2AB}$
C	AAC	$\frac{A \cdot SIN\ C}{SIN\ A}$
C	ABC	$\sqrt{A^2+B^2-(2AB \cdot COS\ C)}$
C	BBC	$\frac{B \cdot SIN\ C}{SIN\ B}$
AREA S=	ABC	$\frac{A \cdot B \cdot SIN\ C}{2}$
S=$\frac{1}{2}$(A+B+C)	A,B,C	$\sqrt{S(S-A)(S-B)(S-C)}$

119

(Continued on next page)

TRIGONOMETRIC SOLUTION OF TRIANGLES (Cont.)

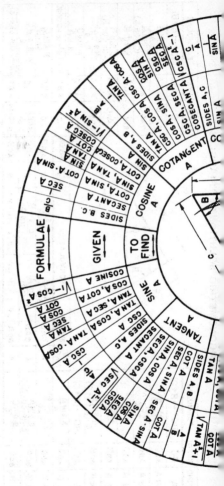

120

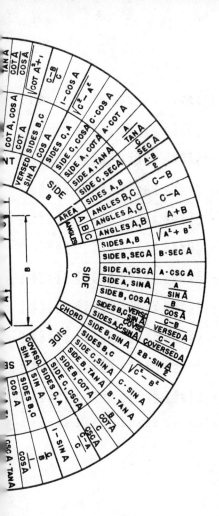

121

FOUR-PLACE LOG TABLES

N	0	1	2	3	4	5	6	7	8	9	n. d.
10	0000	0043	0086	0128	0170	0212	0253	0294	0334	0374	4.2
11	0414	0453	0492	0531	0569	0607	0645	0682	0719	0755	3.8
12	0792	0828	0864	0899	0934	0969	1004	1038	1072	1106	3.5
13	1139	1173	1206	1239	1271	1303	1335	1367	1399	1430	3.2
14	1461	1492	1523	1553	1584	1614	1644	1673	1703	1732	3.0
15	1761	1790	1818	1847	1875	1903	1931	1959	1987	2014	2.8
16	2041	2068	2095	2122	2148	2175	2201	2227	2253	2279	2.6
17	2304	2330	2355	2380	2405	2430	2455	2480	2504	2529	2.5
18	2553	2577	2601	2625	2648	2672	2695	2718	2742	2765	2.4
19	2788	2810	2833	2856	2878	2900	2923	2945	2967	2989	2.2
20	3010	3032	3054	3075	3096	3118	3139	3160	3181	3201	2.1
21	3222	3243	3263	3284	3304	3324	3345	3365	3385	3404	2.0
22	3424	3444	3464	3483	3502	3522	3541	3560	3579	3598	1.9
23	3617	3636	3655	3674	3692	3711	3729	3747	3766	3784	1.8
24	3802	3820	3838	3856	3874	3892	3909	3927	3945	3962	1.8
25	3979	3997	4014	4031	4048	4065	4082	4099	4116	4133	1.7
26	4150	4166	4183	4200	4216	4232	4249	4265	4281	4298	1.6
27	4314	4330	4346	4362	4378	4393	4409	4425	4440	4456	1.6
28	4472	4487	4502	4518	4533	4548	4564	4579	4594	4609	1.5
29	4624	4639	4654	4669	4683	4698	4713	4728	4742	4757	1.5
30	4771	4786	4800	4814	4829	4843	4857	4871	4886	4900	1.4

31	4914	4928	4942	4955	4969	4983	4997	5011	5024	5038	1.4
32	5051	5065	5079	5092	5105	5119	5132	5145	5159	5172	1.3
33	5185	5198	5211	5224	5237	5250	5263	5276	5289	5302	1.3
34	5315	5328	5340	5353	5366	5378	5391	5403	5416	5428	1.3
35	5441	5453	5465	5478	5490	5502	5514	5527	5539	5551	1.2
36	5563	5575	5587	5599	5611	5623	5635	5647	5658	5670	1.2
37	5682	5694	5705	5717	5729	5740	5752	5763	5775	5786	1.2
38	5798	5809	5821	5832	5843	5855	5866	5877	5888	5899	1.1
39	5911	5922	5933	5944	5955	5966	5977	5988	5999	6010	1.1
40	6021	6031	6042	6053	6064	6075	6085	6096	6107	6117	1.1
41	6128	6138	6149	6160	6170	6180	6191	6201	6212	6222	1.0
42	6232	6243	6253	6263	6274	6284	6294	6304	6314	6325	1.0
43	6335	6345	6355	6365	6375	6385	6395	6405	6415	6425	1.0
44	6435	6444	6454	6464	6474	6484	6493	6503	6513	6522	1.0
45	6532	6542	6551	6561	6571	6580	6590	6599	6609	6618	1.0
46	6628	6637	6646	6656	6665	6675	6684	6693	6702	6712	.9
47	6721	6730	6739	6749	6758	6767	6776	6785	6794	6803	.9
48	6812	6821	6830	6839	6848	6857	6866	6875	6884	6893	.9
49	6902	6911	6920	6928	6937	6946	6955	6964	6972	6981	.9
50	6990	6998	7007	7016	7024	7033	7042	7050	7059	7067	.9
51	7076	7084	7093	7101	7110	7118	7126	7135	7143	7152	.8
52	7160	7168	7177	7185	7193	7202	7210	7218	7226	7235	.8
53	7243	7251	7259	7267	7275	7284	7292	7300	7308	7316	.8
54	7324	7332	7340	7348	7356	7364	7372	7380	7388	7396	.8

(Continued on next page)

FOUR-PLACE LOG TABLES (Cont.)

N	0	1	2	3	4	5	6	7	8	9	u. d.
55	7404	7412	7419	7427	7435	7443	7451	7459	7466	7474	.8
56	7482	7490	7497	7505	7513	7520	7528	7536	7543	7551	.8
57	7559	7566	7574	7582	7589	7597	7604	7612	7619	7627	.8
58	7634	7642	7649	7657	7664	7672	7679	7686	7694	7701	.7
59	7709	7716	7723	7731	7738	7745	7752	7760	7767	7774	.7
60	7782	7789	7796	7803	7810	7818	7825	7832	7839	7846	.7
61	7853	7860	7868	7875	7882	7889	7896	7903	7910	7917	.7
62	7924	7931	7938	7945	7952	7959	7966	7973	7980	7987	.7
63	7993	8000	8007	8014	8021	8028	8035	8041	8048	8055	.7
64	8062	8069	8075	8082	8089	8096	8102	8109	8116	8122	.7
65	8129	8136	8142	8149	8156	8162	8169	8176	8182	8189	.7
66	8195	8202	8209	8215	8222	8228	8235	8241	8248	8254	.7
67	8261	8267	8274	8280	8287	8293	8299	8306	8312	8319	.6
68	8325	8331	8338	8344	8351	8357	8363	8370	8376	8382	.6
69	8388	8395	8401	8407	8414	8420	8426	8432	8439	8445	.6
70	8451	8457	8463	8470	8476	8482	8488	8494	8500	8506	.6
71	8513	8519	8525	8531	8537	8543	8549	8555	8561	8567	.6
72	8573	8579	8585	8591	8597	8603	8609	8615	8621	8627	.6
73	8633	8639	8645	8651	8657	8663	8669	8675	8681	8686	.6
74	8692	8698	8704	8710	8716	8722	8727	8733	8739	8745	.6

	0	1	2	3	4	5	6	7	8	9	
75	8751	8756	8762	8768	8774	8779	8785	8791	8797	8802	.6
76	8808	8814	8820	8825	8831	8837	8842	8848	8854	8859	.6
77	8865	8871	8876	8882	8887	8893	8899	8904	8910	8915	.6
78	8921	8927	8932	8938	8943	8949	8954	8960	8965	8971	.6
79	8976	8982	8987	8993	8998	9004	9009	9015	9020	9025	.5
80	9031	9036	9042	9047	9053	9058	9063	9069	9074	9079	.5
81	9085	9090	9096	9101	9106	9112	9117	9122	9128	9133	.5
82	9138	9143	9149	9154	9159	9165	9170	9175	9180	9186	.5
83	9191	9196	9201	9206	9212	9217	9222	9227	9232	9238	.5
84	9243	9248	9253	9258	9263	9269	9274	9279	9284	9289	.5
85	9294	9299	9304	9309	9315	9320	9325	9330	9335	9340	.5
86	9345	9350	9355	9360	9365	9370	9375	9380	9385	9390	.5
87	9395	9400	9405	9410	9415	9420	9425	9430	9435	9440	.5
88	9445	9450	9455	9460	9465	9469	9474	9479	9484	9489	.5
89	9494	9499	9504	9509	9513	9518	9523	9528	9533	9538	.5
90	9542	9547	9552	9557	9562	9566	9571	9576	9581	9586	.5
91	9590	9595	9600	9605	9609	9614	9619	9624	9628	9633	.5
92	9638	9643	9647	9652	9657	9661	9666	9671	9675	9680	.5
93	9685	9689	9694	9699	9703	9708	9713	9717	9722	9727	.5
94	9731	9736	9741	9745	9750	9754	9759	9763	9768	9773	.5
95	9777	9782	9786	9791	9795	9800	9805	9809	9814	9818	.5
96	9823	9827	9832	9836	9841	9845	9850	9854	9859	9863	.4
97	9868	9872	9877	9881	9886	9890	9894	9899	9903	9908	.4
98	9912	9917	9921	9926	9930	9934	9939	9943	9948	9952	.4
99	9956	9961	9965	9969	9974	9978	9983	9987	9991	9996	.4

TWIST DRILL SIZES

Size	Decimal equivalents	Size	Decimal equivalents
1/2	0.5000	E 1/4	.2500
31/64	.4844	D	.2460
15/32	.4687	C	.2420
29/64	.4531	B	.2380
7/16	.4375	15/64	.2344
27/64	.4219	A	.2340
Z	.4130	No. 1	.2280
13/32	.4062	2	.2210
Y	.4040	7/32	.2187
X	.3970	3	.2130
25/64	.3906	4	.2090
W	.3860	5	.2055
V	.3770	6	.2040
3/8	.3750	13/64	.2031
U	.3680	7	.2010
23/64	.3594	8	.1990
T	.3580	9	.1960
S	.3480	10	.1935
11/32	.3437	11	.1910
R	.3390	12	.1890
Q	.3320	3/16	.1875
21/64	.3281	13	.1850
P	.3230	14	.1820
O	.3160	15	.1800
5/16	.3125	16	.1770
N	.3020	17	.1730
19/64	.2969	11/64	.1719
M	.2950	18	.1695
L	.2900	19	.1660
9/32	.2812	20	.1610
K	.2810	21	.1590
J	.2770	22	.1570
I	.2720	5/32	.1562
H	.2660	23	.1540
17/64	.2656	24	.1520
G	.2610	25	.1495
F	.2570	26	.1470
		27	.1440
		9/64	.1406

Size	Decimal equivalents	Size	Decimal equivalents
28	.1405	45	.0820
29	.1360	46	.0810
30	.1285	47	.0785
$\frac{1}{8}$	.1250	$\frac{5}{64}$	.0781
31	.1200	48	.0760
32	.1160	49	.0730
33	.1130	50	.0700
34	.1110	51	.0670
35	.1100	52	.0635
$\frac{7}{64}$	.1094	$\frac{1}{16}$	.0625
36	.1065	53	.0595
37	.1040	54	.0550
38	.1015	55	.0520
39	.0995	$\frac{3}{64}$	.0469
40	.0980	56	.0465
41	.0960	57	.0430
$\frac{3}{32}$	.0937	58	.0420
42	.0935	59	.0410
43	.0890	60	.0400
44	.0860		

THREAD AND TAP DRILL SIZES

Nominal size	Thr'd series	Major diameter inches
0–80	N. F.	.0600
1–64	N. C.	.0730
72	N. C.	.0730
2–56	N. C.	.0860
64	N. F.	.0860
3–18	N. C.	.0990
56	N. F.	.0990
4–40	N. C.	.1120
48	N. F.	.1120
5–40	N. C.	.1250
44	N. F.	.1250
6–32	N. C.	.1380
40	N. F.	.1380
8–32	N. C.	.1640
36	N. F.	.1640
10–24	N. C.	.1900
32	N. F.	.1900
12–24	N. C.	.2160
28	N. F.	.2160
$\frac{1}{4}$–20	N. C.	.2500
28	N. F.	.2500
$\frac{5}{16}$–18	N. C.	.3125
24	N. F.	.3125
$\frac{3}{8}$–16	N. C.	.3750
24	N. F.	.3750
$\frac{7}{16}$–14	N. C.	.4375
20	N. F.	.4375
$\frac{1}{2}$–13	N. C.	.5000
20	N. F.	.5000
$\frac{9}{16}$–12	N. C.	.5625
18	N. F.	.5625
$\frac{5}{8}$–11	N. C.	.6250
18	N. F.	.6250
$\frac{3}{4}$–10	N. C.	.7500
16	N. F.	.7500
$\frac{7}{8}$– 9	N. C.	.8750
14	N. F.	.8750
1– 8	N. C.	1.0000
14	N. F.	1.0000

Root diameter, inches	Tap drill to produce approx. 75 % full thread	Decimal equivalent of tap drill
.0438	$\frac{3}{64}$	.0469
.0527	53	.0595
.0550	53	.0595
.0628	50	.0700
.0657	50	.0700
.0719	47	.0785
.0758	45	.0820
.0795	43	.0890
.0849	42	.0935
.0925	38	.1015
.0955	37	.1040
.0974	36	.1065
.1055	33	.1130
.1234	29	.1360
.1279	29	.1360
.1359	25	.1495
.1494	21	.1590
.1619	16	.1770
.1696	14	.1820
.1850	7	.2010
.2036	3	.2130
.2403	F	.2570
.2584	I	.2720
.2938	$\frac{5}{16}$	.3125
.3209	Q	.3320
.3447	U	.3680
.3726	$\frac{25}{64}$	.3906
.4001	$\frac{27}{64}$	.4219
.4351	$\frac{29}{64}$	.4531
.4542	$\frac{31}{64}$	.4844
.4903	$\frac{33}{64}$	.5156
.5069	$\frac{17}{32}$	.5312
.5528	$\frac{37}{64}$	.5781
.6201	$\frac{21}{32}$	.6562
.6688	$\frac{11}{16}$	.6875
.7307	$\frac{49}{64}$	.7656
.7822	$\frac{13}{16}$	.8125
.8376	$\frac{7}{8}$	.8750
.9072	$\frac{15}{16}$	.9375

COMMON DC BRIDGES

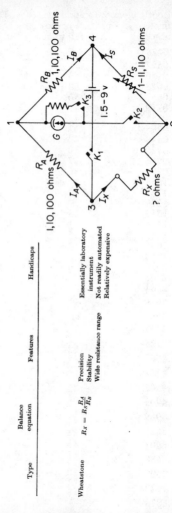

Type	Balance equation	Features	Handicaps
Wheatstone	$R_X = R_S \frac{R_A}{R_B}$	Precision Stability Wide resistance range	Essentially laboratory instrument Not readily automated Relatively expensive

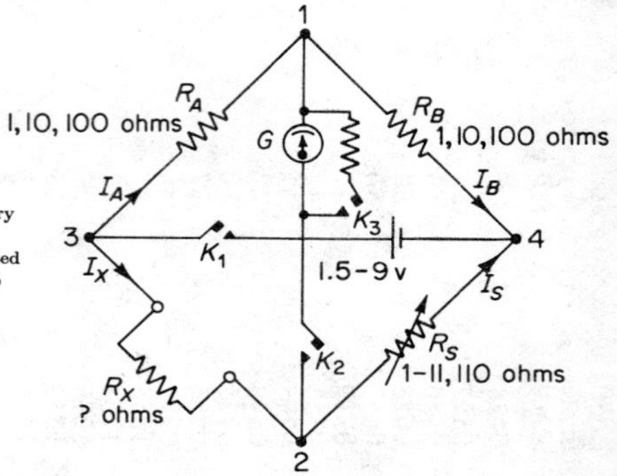

1
2
3
4
R_A
1, 10, 100 ohms
R_B
1, 10, 100 ohms
I_A
I_B
I_X
I_S
G
K_1
K_2
K_3
1.5–9 v
R_S
1–11, 110 ohms
R_X
? ohms

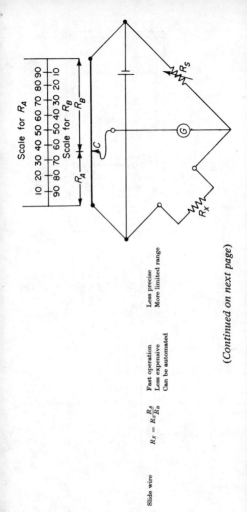

Scale for R_A

Scale for R_B

R_A R_B

| 10 | 20 | 30 | 40 | 50 | 60 | 70 | 80 | 90 |
| 90 | 80 | 70 | 60 | 50 | 40 | 30 | 20 | 10 |

Slide wire

$$R_X = R_S \frac{R_A}{R_B}$$

Fast operation
Less expensive
Can be automated

Less precise
More limited range

(Continued on next page)

131

COMMON DC BRIDGES (Cont.)

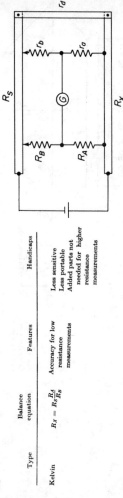

Type	Balance equation	Features	Handicaps
Kelvin	$R_X = R_S \dfrac{R_A}{R_B}$	Accuracy for low resistance measurements	Less sensitive Less portable Added parts not needed for higher resistance measurements

132

COMMON AC BRIDGES

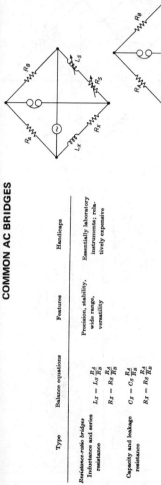

Type	Balance equations	Features	Handicaps
Resistance-ratio bridges			
Inductance and series resistance	$L_X = L_S \dfrac{R_A}{R_B}$ $R_X = R_S \dfrac{R_A}{R_B}$	Precision, stability, wide range, versatility	Essentially laboratory instruments; relatively expensive
Capacity and leakage resistance	$C_X = C_S \dfrac{R_A}{R_B}$ $R_X = R_S \dfrac{R_A}{R_B}$		

(Continued on next page)

133

COMMON AC BRIDGES (Cont.)

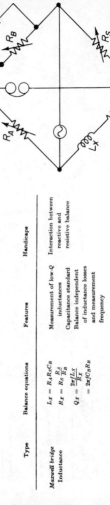

Type	Balance equations	Features	Handicaps
Maxwell bridge Inductance	$L_X = R_A R_S C_B$ $R_X = R_S \dfrac{R_A}{R_B}$ $Q_X = \dfrac{2\pi f L_X}{R_X}$ $= 2\pi f C_B R_B$	Measurement of low-Q inductances Capacitance standard Balance independent of inductance losses and measurement frequency	Interaction between reactive and resistive balance

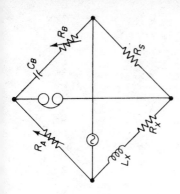

Hay bridge
Inductance

$$L_X = \frac{R_A R_S C_B}{1 + (2\pi f C_B R_B)^2}\text{F}$$

$$= \frac{R_A R_S C_B}{1 + (1/Q_X)^2}$$

$$R_X = R_S \frac{R_A}{R_B} \cdot \frac{2\pi f L_X}{R_X}$$

$$Q_X = \frac{2\pi f L_X}{R_X}$$

$$= \frac{1}{2\pi f C_B R_B}$$

Measurement of
high-Q inductances
Capacitance standard

Inductance measure-
ment dependent
upon Q and
measurement
frequency
Complex balance
equations

(Continued on next page)

135

COMMON AC BRIDGES (Cont.)

Type	Balance equations	Features	Handicaps
Schering bridge Capacity, dissipation factor	$C_X = C_A \dfrac{R_B}{R_S}$ $R_X = R_S \dfrac{C_B}{C_A}$ $Q_X = \dfrac{1}{2\pi f C_X R_X}$ $ = \dfrac{1}{2\pi f C_B R_B}$	Accurate capacitance measurement Measures dissipation factor of capacitors	Essentially laboratory instrument

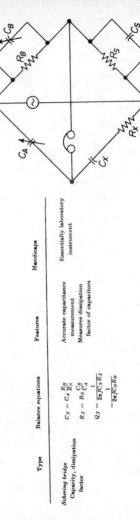

Wien bridge
Capacity, frequency

$$\frac{C_X}{C_S} = \frac{R_B}{R_A} - \frac{R_S}{R_X}$$

$$(2\pi f)^2 = \frac{1}{R_X R_S C_X C_S}$$

$$C_X^2 = \frac{R_B R_X - R_A R_S}{(2\pi f)^2 R_A R_S R_X^2}$$

$$C_S^2 = \frac{R_A}{R_S(2\pi f)^2 (R_B R_X - R_A R_S)}$$

Essentially laboratory instrument
Complex balance equations

Accurate capacitance measurement
Radio-frequency measurement
Readily derived from resistance-capacity bridge

SUMMARY OF FILTER CIRCUIT CHARACTERISTICS

TYPE	ATTENUATION	STRUCTURE π	STRUCTURE T	ELEMENT C_1	ELEMENT L_1
"K"	(graph, f_c)	L_1; $\frac{C_2}{2}$ $\frac{C_2}{2}$	$L_1/2$ $L_1/2$; C_2		$\dfrac{Z_0}{\pi f_c}$ $-L_k-$
"m_1"	(graph, f_c f_∞)	L_1; $2L_2$ $2L_2$; $\frac{C_2}{2}$ $\frac{C_2}{2}$	$L_1/2$ $L_1/2$; L_2; C_2		mL_k
"m_2"	(graph, f_c f_∞)	C_1; L_1; $\frac{C_2}{2}$ $\frac{C_2}{2}$	$2C_1$ $2C_1$; $L_1/2$ $L_1/2$; C_2	$\dfrac{1-m^2}{4m}C_k$	mL_k
"K"	(graph, f_c)	C_1; $2L_2$ $2L_2$	$2C_1$ $2C_1$; L_2	$\dfrac{1}{4\pi f_c Z_0}$ $-C_k-$	
"m_1"	(graph, f_∞ f_c)	$2L_2$ $2L_2$; C_1; $\frac{C_2}{2}$ $\frac{C_2}{2}$	$2C_1$ $2C_1$; L_2; C_2	$\dfrac{C_k}{m}$	
"m_2"	(graph, f_∞ f_c)	C_1; L_1; $2L_2$ $2L_2$	$2C_1$ $2C_1$; $L_1/2$ $L_1/2$; L_2	$\dfrac{C_k}{m}$	$\dfrac{4m}{1-m^2}L$

CONSTANTS		CHARACTERISTIC POINTS			
C_2	L_2	PASSBAND Z_0	$f\infty$	f_c	$\dfrac{f\infty}{f_c}$
$\dfrac{1}{\pi f_c Z_0}$ -C_k-		$\sqrt{\dfrac{L_1}{C_2}}$	∞	$\dfrac{1}{\pi\sqrt{L_1 C_2}}$	∞
mC_k	$\dfrac{1-m^2}{4m}L_k$	$\sqrt{\dfrac{L_1}{C_2}}$	$\dfrac{1}{2\pi\sqrt{L_2 C_2}}$	$\dfrac{1}{\pi\sqrt{C_2(L_1+4L_2)}}$	$\sqrt{1+\dfrac{L_1}{4L_2}}$
mC_k		$\sqrt{\dfrac{L_1}{C_2}}$	$\dfrac{1}{2\pi\sqrt{L_1 C_1}}$	$\dfrac{1}{\pi\sqrt{L_1(C_2+4C_1)}}$	$\sqrt{1+\dfrac{C_2}{4C_1}}$
	$\dfrac{Z_0}{4\pi f_c}$ -L_k-	$\sqrt{\dfrac{L_2}{C_1}}$	0	$\dfrac{1}{4\pi\sqrt{L_2 C_1}}$	∞
$\dfrac{4m}{1-m^2}C_k$	$\dfrac{L_k}{m}$	$\sqrt{\dfrac{L_2}{C_1}}$	$\dfrac{1}{2\pi\sqrt{L_2 C_2}}$	$\dfrac{1}{4\pi}\sqrt{\dfrac{1}{L_2 C_1}+\dfrac{4}{L_2 C_2}}$	$\sqrt{1+\dfrac{C}{4C_1}}$
	$\dfrac{L_k}{m}$	$\sqrt{\dfrac{L_2}{C_1}}$	$\dfrac{1}{2\pi\sqrt{L_1 C_1}}$	$\dfrac{1}{4\pi}\sqrt{\dfrac{1}{L_2 C_2}+\dfrac{4}{L_1 C_1}}$	$\sqrt{1+\dfrac{L_1}{4L_2}}$

FILTER CIRCUIT DESIGN NOMOGRAPH

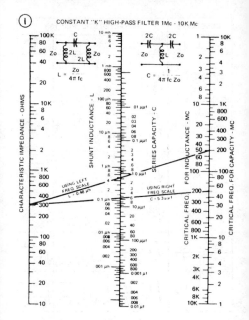

① CONSTANT "K" HIGH-PASS FILTER 1Mc - 10K Mc

$$L = \frac{Zo}{4\pi fc}$$

$$C = \frac{1}{4\pi fc\ Zo}$$

CHARACTERISTIC IMPEDANCE - OHMS

SHUNT INDUCTANCE - L

SERIES CAPACITY - C

CRITICAL FREQ. FOR INDUCTANCE - MC

CRITICAL FREQ. FOR CAPACITY - MC

USING LEFT FREQ SCALE
L = 0.48 μh

USING RIGHT FREQ SCALE
C = 5.3 μμf

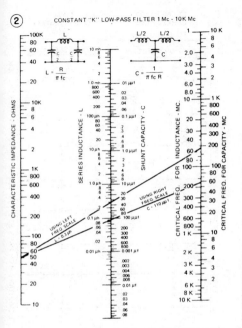

CONSTANT "K" LOW-PASS FILTER 1 Mc - 10K Mc

Reprinted from *Electronic Industries*, A Chilton publication, Feb. 1954 © Chilton Co.

ELEMENTARY FILTER CIRCUITS

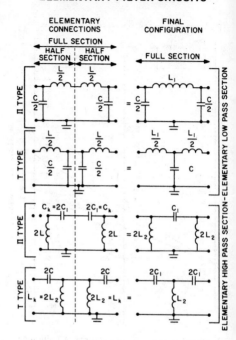

CIRCUIT Q—ITS NATURE AND MEASUREMENT

The letter Q when used in describing the qualities of a resonant circuit is widespread in its application to almost every phase of performance. Basically, it expresses a comparison, during one cycle of operation, giving the ratio of the total energy stored in a circuit versus the energy dissipated.

From Q measurements we can determine:

1. The damping effect when current is decaying in a resonant circuit.

2. Phase angle and power factor of tuned circuits.

3. Selectivity of a tuned circuit.

4. Radio-frequency resistance of a coil.

5. The loss angle of a capacitor at RF.

6. Dielectric constants.

7. Antenna characteristics.

8. Transmission-line parameters.

The Q-Meter operating in a circuit measures the resonant rise of voltage across either of the reactive elements in a tuned circuit. This measurement results from injecting a small, known radio-frequency voltage across a very small series resistor which is part of the resonant circuit. The magnified or resonant voltage rise is measured by a vacuum-tube voltmeter. Since the amount of injected voltage is accurately known, the resonant voltage rise or the circuit magnification factor can be directly calibrated in terms of the Q of the coil being measured.

Q *in series resonant circuits* is calculated using the effective series resistance (R_s), which is a fictitious value representing all losses including those of the resonant coil and capacitor. Algebraically, using R_s, we express it:

$$Q = \frac{\omega L}{R_s} = \frac{1}{\omega C R_s}$$

(Continued on next page)

CIRCUIT Q (Cont.)

Thus we see $R_s = \omega L/Q = 1/Q\omega C$ and is usually a very *small* quantity.

Q in a parallel resonant circuit is calculated using the effective parallel resistance (R_p) which is a reflected fictitious value representing all losses (including those of the resonant coil and capacitor) and which is represented by a parallel resistor connected *across* the turned circuit. We express it:

$$Q = \frac{R_p}{\omega L} = \omega C R_p$$

Again we see $R_p = Q\omega L = Q/\omega C$ and represents a very *large* quantity.

Here we see that the impedance of a parallel resonant circuit is Q times the impedance of the reactive elements. Also, in the series case, the current flowing at resonance is Q times the normal current flow.

Q in a damped oscillating circuit tells us the logarithmic decrement δ of the circuit. Here:

$$\delta = \frac{\pi}{Q} = \frac{R_s}{2fL} = 2\pi^2 fCR_s$$

Now, $R/2L$ is the damping coefficient which tells the amount by which the amplitude of each successive cycle is lower than its predecessor. So δ, which multiplies this number by f, accounts for frequency and is logarithmic in nature since the equation for current decay is:

$$\frac{I_2}{I_1} = \epsilon^{-RT/2L}$$

Q and circuit phase angle. The vector relationship between the current and driving voltage in a resonant circuit is the familiar formula:

$$\tan\phi = \frac{\omega L}{R_s} = Q$$

where ϕ = the phase angle.

Q and power factor. Of similar nature is the power factor of an inductor or the ratio of the total effective resistance to the total circuit impedance.

$$Q = \frac{1}{\cos\phi}$$

144

FUSE DATA

Table 1 Characteristics of fuses in electronic applications

Blowing characteristics	Fuse type	Physical size	Ampere ratings	Max. voltage rating
High speed	8AG	1" × 1/4" diam.	1/500-5	32, 125, 250
Medium lag	3AG	1-1/4" × 1/4" diam.	1/16-20	32, 125, 250
	5AG	1-1/2" × 13/32" diam.	1-50	32, 250
Slo-blo	3AG	1-1/4" × 1/4" diam.	1/100-5	32, 125
	5AG	1-1/2" × 13/32" diam.	1-30	32, 125

Table 2 Fuse sizes for protection of insulated copper wire

Wire size (B&S)	Fuse rating (amp)
No. 16	10
No. 14	15
No. 12	20-30
No. 10	40
No. 8	50

Table 3 Medium size lag fuses for 115-v power supplies

Power supply rating (watts)	Fuse rating (amp)
40-65	1
65-100	$1\frac{1}{2}$
100-150	2
150-250	3
250-350	5
350-450	6

(Continued on next page)

145

FUSE DATA (Cont.)

Table 4   Current ratings of fuses for various instrument ranges

Fuse rating (amp)	Max. load (ma)	Voltmeters (ohms per volt)	Milliammeters (all magnetic-type movements)	Milliammeters (thermocouple types)
1/500	2	Over 1,000	Galvanometers	0-0.1 to 0-0.5
1/200	5	Over 1,000	Galvanometers	Up to 0.1
1/100	10	1,000	Up to 0.1	0-5 to 0-10
1/32	25	500-1,000	0-0.1 to 0-10	0-10 to 0-25
1/16	60	100-500	0-10 to 0-25	0-25 to 0-60
1/8	100	20-100	0-25 to 0-75	0-75 to 0-115
1/4	200	10-20	0-75 to 0-150	0-115 to 0-200
3/8	300	5-10	0-150 to 0-250	0-200 to 0-300
1/2	400	3-5	0-250 to 0-350	0-300 to 0-400
3/4	600		0-350 to 0-500	0-400 to 0-600
1	1,000		0-500 to 0-750	0-600 to 0-1,000
1½	1,500		0-750 to 0-1,000	0-1,000 to 0-1,500
2	2,000		0-1,000 to 0-1,500	0-1,500 to 0-2,000
3	3,000		0-1,500 to 0-2,000	0-2,000 to 0-3,000
5	5,000		0-2,000 to 0-4,000	0-3,000 to 0-5,000

146

Table 5 Fuse wire table

Fusing current, amps.	Copper		Tin		Allo-Tin		Lead	
	Diam., inch	S.W.G. (approx.)	Diam., inch	S.W.G. (approx.)	Diam., inch	S.W.G. (approx.)	Diam., inch	S.W.G. (approx.)
1	.0021	47	.0072	37	.0083	35	.0081	35
2	.0034	43	.0113	31	.0132	29	.0128	30
3	.0044	41	.0149	28	.0173	27	.0168	27
4	.0053	39	.0181	26	.0210	25	.0203	25
5	.0062	38	.0210	25	.0243	23	.0236	23
10	.0098	33	.0334	21	.0386	19	.0375	20
15	.0129	30	.0437	19	.0506	18	.0491	18
20	.0156	28	.0529	17	.0613	16	.0595	17
25	.0181	26	.0614	16	.0711	15	.0690	15
30	.0205	25	.0694	15	.0803	14	.0779	14
40	.0248	23	.0840	14	.0973	13	.0944	13
50	.0288	22	.0975	13	.1129	11	.1095	12
70	.0360	20	.1220	10	.1413	9	.1371	9
100	.0457	18	.1548	8	.1792	7	.1739	7

BASIC RECTIFIER CIRCUITS

Type of circuit

Primary

Secondary

One cycle wave of rectifier output voltage
(no overlap)

Number of rectifier elements in circuit

RMS DC volts output
Peak DC volts output
Peak reverse volts per rectifier element

Average DC output current
Average DC output current per rectifier element
RMS current per rectifier element {Resistive load / Inductive load

Peak current per rectifier element {Resistive load / Inductive load

Ratio: Peak to average current {Resistive load / Inductive load
per element

% ripple $\left(\dfrac{\text{RMS of ripple}}{\text{average output voltage}}\right)$

Transformer secondary RMS volts per leg

Transformer secondary RMS volts line-to-line
Secondary line current
Transformer secondary volt-amperes per leg
Transformer primary RMS amperes per leg
Transformer primary volt-amperes per leg
Average of primary and secondary volt-amperes
Primary line current
Line power factor

Single phase half-wave	Single phase center-tap	Single phase bridge
1	2	4
1.57	1.11	1.11
3.14	1.57	1.57
3.14	3.14	1.57
1.41	2.82	1.41
1.41	1.41	1.41
1.00	1.00	1.00
1.00	0.500	0.500
1.57	0.785	0.785
....	0.707	0.707
3.14	1.57	1.57
....	1.00	1.00
3.14	3.14	3.14
....	2.00	2.00
121%	48%	48%
Resistive load		Inductive
2.22	1.11 (To center-tap)	1.11 (Total)
2.22	2.22	1.11
1.57	0.707	1.00
3.49	1.57	1.11
1.57	1.00	1.00
3.49	1.11	1.11
3.49	1.34	1.11
1.57	1.00	1.00
....	0.900	0.900

VACUUM TUBE AMPLIFIER CIRCUITS

(*a*) RESISTANCE-CAPACITANCE COUPLED has simple, economical construction, and wide frequency range.

$$\frac{\text{Stage}}{\text{Gain}} = \frac{V_{\text{out}}}{V_{\text{in}}} = \mu \times \frac{R_L}{R_L + P_P}$$

(*b*) TRANSFORMER COUPLED has no coupling capacitor, uses full +B voltage on plates; receives step-up from transformer, restricts frequency range.

$$\frac{\text{Stage}}{\text{Gain}} = \frac{V_{\text{out}}}{V_{\text{in}}} = \frac{N_2}{N_1} \times \frac{X_2}{X_L + R_P}$$

(*c*) IMPEDANCE COUPLED has single winding plus coupling capacitor. Restricts frequency range; full +B voltage on plates.

$$\frac{\text{Stage}}{\text{Gain}} = \frac{V_{\text{out}}}{V_{\text{in}}} = \mu \times \frac{X_L}{X_L + R_P}$$

(*d*) TUNED CIRCUIT COUPLED has high gain, good selectivity, tunes out grid and plate capacities.

$$\frac{\text{Stage}}{\text{Gain}} = \frac{V_{\text{out}}}{V_{\text{in}}} = G_M \times \frac{L}{CR}$$

(*e*) DIRECT COUPLED has no coupling capacitor nor grid and cathode resistors, uses large, expensive voltage divider, bypassing capacitors. State gain same as resistance coupled stage, with wide frequency range.

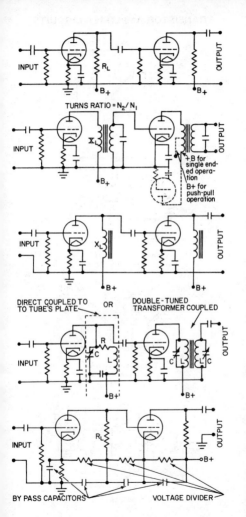

INPUT

OUTPUT

R_L

B+

B+

TURNS RATIO = N_2/N_1

X_L

INPUT

OUTPUT

B+

+B for single ended operation

B+ for push-pull operation

INPUT

X_L

OUTPUT

B+

B+

DIRECT COUPLED TO TUBE'S PLATE   OR   DOUBLE-TUNED TRANSFORMER COUPLED

R

C

L

INPUT

C

C

L

OUTPUT

B+

B+

INPUT

R_L

OUTPUT

B+

BY PASS CAPACITORS   VOLTAGE DIVIDER

151

TRANSISTOR AMPLIFIER CIRCUITS

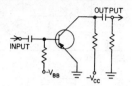

Resistance-capacitance (R-C) coupled

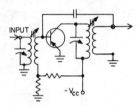

*Transformer-coupled
(neutralized video I F stage)*

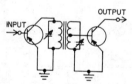

*Transformer-coupled
(double tuned circuit)*

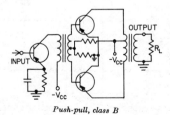

Push-pull, class B

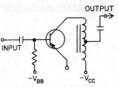

Impedance-coupled (video)

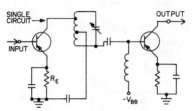

Impedance-coupled (tuned circuit)

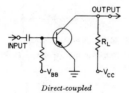

Direct-coupled

VACUUM TUBE OSCILLATOR CIRCUITS

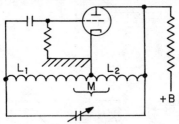

(A) HARTLEY OSCILLATOR has inductive feedback due to coupling between L_1 and L_2.

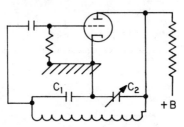

(B) COLPITTS OSCILLATOR has capacitive feedback due to voltage across C_1.

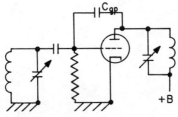

(C) TUNED GRID-TUNED PLATE OSCILLATOR derives feedback to grid from plate through grid-plate capacity C.

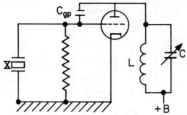

(D) CRYSTAL OSCILLATOR utilizes resonant properties of crystal X, derives feedback from resonant circuit LC through C.

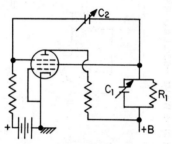

(E) R-C OSCILLATOR has feedback from $C_1 R_1$ network; through C_2 utilizes negative resistance developed in screen circuit to maintain oscillation; must operate at low voltages; has better frequency range than L-C oscillators.

TRANSISTOR SINE WAVE OSCILLATOR CIRCUITS

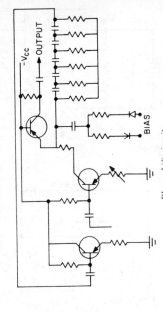

Phase-shift circuit

Push-pull circuit

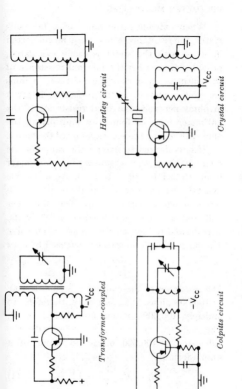

Hartley circuit

Crystal circuit

Transformer-coupled

Colpitts circuit

CALCULATING METER MULTIPLIERS AND SHUNTS

VOLTMETER MULTIPLIERS

When extending the range of a DC voltmeter, the resistance which must be connected in series with the meter is easily calculated, provided that either the internal resistance of the voltmeter in ohms or the resistance in ohms per volt is known. If the resistance is given in ohms per volt, the total resistance of the voltmeter may be found by multiplying the ohms per volt by the scale reading of the meter.

If it is desired to extend the range of the meter 10 times, the resistance of the voltmeter is multiplied by $10 - 1$, or 9. As a specific example, let us assume that the voltmeter in question has a sensitivity of 1000 ohms per volt and a full scale deflection of 100 volts. It is desired to increase the range to 500 volts. The range, therefore, is to be increased by 5, or

$$\frac{(500)}{(100)}$$

The resistance of the voltmeter is 1000×100, or 100,000 ohms. It is necessary, therefore, to multiply 100,000 by one less than 5, or 4. The resistance necessary in series with the voltmeter is 400,000 ohms. In terms of a simple formula,

$$R_s = R_m \left[\frac{V_2}{V_1} - 1 \right] \qquad (1)$$

where R_m is the resistance of the voltmeter in ohms, or the number of ohms per volt times the maximum scale reading of the voltmeter prior to the change, V_1 is the original range of the voltmeter, V_2 is the new maximum range desired, and R_s is the fixed external resistor which must be connected in the circuit.

Figure 1 shows a switching arrangement providing two extension ranges for the voltmeter mentioned above: With the 100-volt range available at the number 1 position of the switch, we may extend to a 500-volt range at the number 2 position with a multiplier resistance calculated by using formula (1).

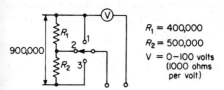

$R_1 = 400,000$
$R_2 = 500,000$
$V = 0$–100 volts
(1000 ohms per volt)

Fig. 1   SIMPLE MULTIPLIER SWITCHING

Thus,

$$R_s = 400,000 \text{ ohms}$$

where $R_m = 100,000$, $V_2 = 500$, and $V_1 = 100$.

In a similar manner, we arrive at a 900,000 ohms multiplier for a 1000-volt range at position number 3. Since we already have a resistance of 400,000 ohms for the 500-volt range, we merely add 500,000 ohms in series to obtain our second multiplier value. An accuracy of 1% is generally satisfactory for these resistors, unless a very high precision meter is used. The general types of meters encountered in service work are only accurate to ±2%, so that it is useless and wasteful to use a resistor with an accuracy of better than ±1%.

CALCULATING METER MULTIPLIERS AND SHUNTS (cont'd.)

EXTENDING MILLIAMMETER RANGES.

To extend the range of a DC milliammeter when the resistance of the milliammeter is known, the shunt resistor which must be connected across the terminals of the meter is calculated very simply by dividing the resistance of the meter by $(K - 1)$, where K is equal to the ratio of the desired maximum reading to the original reading of the meter. This is given as

$$R_{\text{sh}} = \frac{R_m}{(K - 1)} \qquad (2)$$

where R_m = resistance of meter, $K = I_2$ = range desired in milliamperes/I = original range in milliamperes, and R_{sh} = value of shunt resistor.

MEASURING METER RESISTANCES.

Unless great care is exercised, it is inadvisable to measure meter movements by bridge or ohmmeter methods. It is preferable to obtain unknown meter resistances by the half deflection method:

Fig. 2   MEASURING METER RESISTANCE

Referring to Fig. 2, a variable high resistance R_1 is connected in series with the meter, and the meter adjusted to exactly full scale deflection. R_2 is then connected to the circuit and adjusted to make the meter read half scale. R_2 is then equal to the meter resistance and may be measured by any of the usual methods.

SWITCHING IN MULTI-RANGE MILLIAMMETERS

It will usually be found advisable, when making a multi-range milliammeter, to increase the meter resistance 5 to 10 times. This may be done by connecting a series resistor external to the meter. The shunt is then figured, using as the meter resistance the combined resistances of the meter and the series resistor. The series resistor serves two purposes. First, it allows the shunt to be of more reasonable value, thus decreasing errors due to contact resistance or to slight miscalculation. Second, in case of momentary overload, the resistor acts as a ballast slowing down the meter action and in many cases saving a meter which might otherwise be ruined.

UNIVERSAL SHUNT SWITCHING

Probably the best multi-range milliammeter circuit is a modification of the "universal shunt" type. This is shown in Fig. 3. This circuit has several advantages over the usual circuit. First, contact resistance of the switch has absolutely no effect on the accuracy of the meter. The usual circuit has the contact resistance in series with the shunt and thus makes the total shunt resistance inaccurate. If the switch contact should happen to be defective in the usual circuit, the meter would be ruined. Second, with the usual circuit, the switch cannot be operated while the meter is in the circuit. With the universal shunt arrangement, the switch may be operated at any time without damage to the meter.

In calculating shunt resistance, we add the series resistor (Fig. 3) to bring the meter plus series resistance to a value of approximately 200 ohms. The total shunt resistance is now figured by formula (2) to make the meter read full scale for the first desired range (5 or 10 milliamperes). The other resistances are figured by formula (3) derived in Fig. 3:

$$R_x + \frac{A + B}{K} \qquad (3)$$

where $A = R_1 + R_2 + R_3$ (the total shunt resistance);

$B = R_4$ (the external series resistor);

and $K =$ the desired range divided by the fundamental range of the meter.

As an example, assume that we have a 0–1 milliampere meter of 50 ohms resistance, and that we want a multi-range meter giving ranges of 0–5, 0–50, and 0–250 milliamperes. Referring to Fig. I-3, we first add R_4 (150 ohms) to the meter resistance thus bringing total B, $(R_m + R_4)$ resistance to 200 ohms. Our next step is to find the total shunt resistance. For

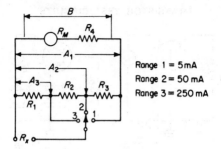

Fig. 3   MULTI-RANGE MILLIAMMETER SWITCHING

Range 1 = 5 mA
Range 2 = 50 mA
Range 3 = 250 mA

this we will use formula (1) and solve for the lowest desired range. For a 5 milliampere range, we would have

$$R_x = A_1 = \frac{200}{5-1} = 50$$

Now we figure the next range using formula (3). For a range of 50 milliamperes, we will have

$$R_x = A_2 = \frac{50 + 200}{50} = 5$$

Since $R_1 + R_2 + R_3 = 50$, R_3 will be 45 ohms. Using the same formula for the 0–250 milliampere range, we have

$$R_x = A_3 = \frac{50 + 200}{250} = 1$$

Since $R_1 + R_2$ was shown to be 5 ohms, R_1 is 1 ohm and R_2 equals 4 ohms.

This method has one advantage over the ordinary calculations: division is usually made by an even number rather than by odd numbers such as 49 and 249, as would be necessary in the usual shunt formulas.

TRANSISTOR TEST CIRCUITS

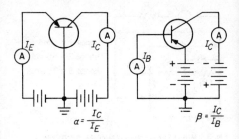

$$\alpha = \frac{I_C}{I_E}$$

$$\beta = \frac{I_C}{I_B}$$

ALPHA AND BETA

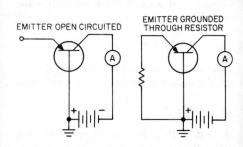

EMITTER OPEN CIRCUITED

EMITTER GROUNDED THROUGH RESISTOR

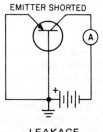

EMITTER SHORTED

LEAKAGE

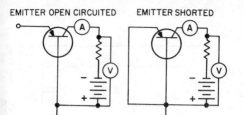

EMITTER OPEN CIRCUITED EMITTER SHORTED

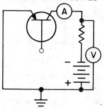

BASE OPEN CIRCUITED

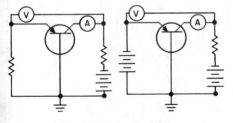

BREAKDOWN VOLTAGE

CIRCUIT FOR TRANSISTOR TESTER

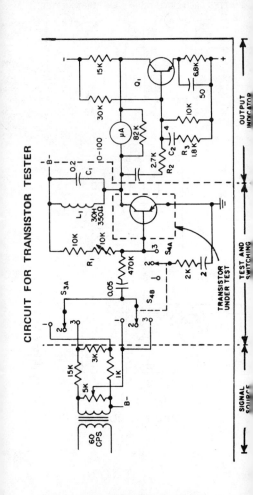

166

TYPES OF RADAR SCANS

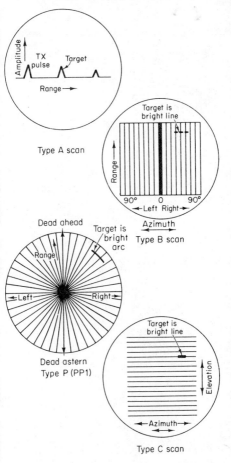

Type A scan

Type B scan

Type P (PP1)

Type C scan

(Continued on next page)

167

TYPES OF RADAR SCANS (Cont.)

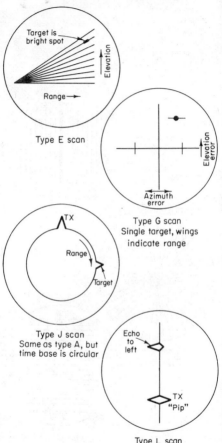

Type E scan

Type G scan
Single target, wings
indicate range

Type J scan
Same as type A, but
time base is circular

Type L scan
Relative strength of pips
on each indicate azimuth

168

BASIC METER MOVEMENTS

Direct Current

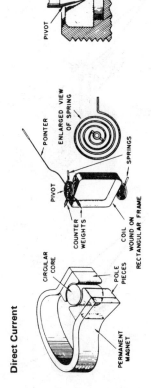

(Continued on next page)

169

BASIC METER MOVEMENTS (Cont.)

Alternating Current

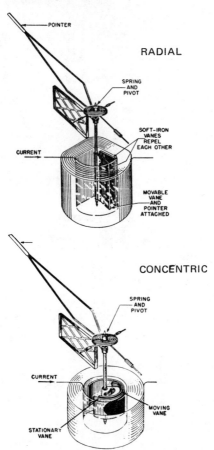

POINTER

RADIAL

SPRING AND PIVOT

SOFT-IRON VANES REPEL EACH OTHER

CURRENT

MOVABLE VANE AND POINTER ATTACHED

CONCENTRIC

SPRING AND PIVOT

CURRENT

MOVING VANE

STATIONARY VANE

MIL DESIGNATION FOR RESISTORS

Description of resistor	Component designation	Range		Specification
		Min. ohms	Max. ohms	
Fixed, composition (insulated)	RC	2.0	30 Meg	MIL-R-11A
Fixed, (composition film, very high frequency)	RF	2.0	1 Meg	MIL-R-10683A
Fixed, film (high stability)	RN	1.0	10 Meg	MIL-R-10509A
Fixed, wirewound (accurate)	RF	2.0	1 Meg	MIL-R-93A
Fixed, wirewound (low power)	RU	.2	2 K	JAN-R-184
Fixed, wirewound (power type)	RW	0	200 K	MIL-R-26B
Variable, composition	RV	100	6 Meg	MIL-R-94-A
Variable, wirewound (low operating temperature)	PA	4.	12 K	JAN-R-19
Variable (wirewound power type)	RP	3.	10 K	MIL-R-22A

MIL DESIGNATION FOR CAPACITORS

Description	Military style designation	Capacity range		Max working volts	Applicable MIL specification
		Min	Max		
Paper	CN	1,000 MMF	2 MF	1,000	MIL-C-91A
Paper	CP	0.001 MF	20 MF	12,500	MIL-C-25A
Mica	CB	10 MMF	3,900 MMF	300	MIL-C-10950A
Mica	CM	5 MMF	0.1 MF	35,000	MIL-C-5A
Ceramic	CV	1.5 MMF	125 MMF	500	JAN-C-81
Ceramic	CC	0.5 MMF	1,300 MMF	500	JAN-C-20A
Ceramic	CK	100 MMF	10,000 MMF	500	MIL-C-11015A
Glass	CY	5 MMF	5,100 MMF	500	MIL-C-11272A
Air	CT	3.5 MMF	143 MMF	†	JAN-C-92
Tantalum	CW	0.5 MF	320 MF	630	[1]MIL-C-18211
Elecrolytic, AC	CJ	21 MF	552 MF	125	[2]MIL-C-25102
Aluminum, Electrolytic	CE	4 MF	2,000 MF	450	[3]MIL-C-14006
	*	0.01 MF	0.25 MF	†	JAN-C-62
	#	0.01 MF	2.0 MF	†	

* Paper, high-voltage, suppression.   † Not specified.   ² USAF.
# Paper, feed-through, suppression.   ¹ Navy.   ³ Sig. Corps.

MIL SPECIFICATIONS FOR
WIRE AND CABLE

MIL-W-76A General specification.

MIL-W-583A Electrical magnet wire.

MIL-W-3861 Solid, bunch-stranded, concentric-lay-stranded, and rope-lay-stranded round electrical wire.

MIL-W-5086A 600 volt single-conductor insulated copper wire for aircraft electrical use.

MIL-W-5274A Aircraft wire—600 volts.

MIL-W-6370B Insulated wire for use in external aircraft antennas.

MIL-W-7072A(ASG) Aluminum wire, 600-volt, insulated, single-conductor, for aircraft electrical power distribution systems.

MIL-W-7139A 600-volt, single-conductor, insulated, copper wire for 400°F, and to operate for a short time in the event of a fire. For airframe wiring.

MIL-W-8777A(ASG) Aircraft wire resistant to flame, fuels, abrasions, fungus, and solvents, with an operating temperature range from −55°C to +150°C at 600 volts.

MIL-W-16878C(Navy) For the internal wiring of meters, panels, and electrical and electronic equipment with minimum size and weight consistent with service requirements. The temperature rating of wire ranges from 80°C to 200°C with potential ratings from 75 to 3,000 volts rms.

MIL-W-19583(Navy) High-temperature magnet wire, three temperature ranges—130°C, 180°C and 200°C. AWG sizes from 4 to 46.

MIL-C-17B Flexible shielded cables, employing solid and semisolid dielectrics. For use as radio-frequency transmission lines—covers coaxial, twin conductor, and twin lead.

MIL-C-3432A Light-, intermediate-, and heavy-duty, flexible and extra-flexible, single-conductor and multiconductor cable, shielded and unshielded, for use in circuits of 300 and 600 volts rms. Also covers heavy-duty, multiconductor, unshielded cable containing ground wires.

(Continued on next page)

MIL SPECIFICATIONS FOR
WIRE AND CABLE (Cont.)

MIL-C-3883 Single-wire and multiwire electrical cord, in which wires are all of the same wire size for use in audio-frequency applications. This is a braided armored cable; its applications include shipboard use.

MIL-C-5756B(ASG) 600-volt heavy-duty, single-conductor wire and multiconductor electric cable for severe flexing service.

MIL-C-7078A This specification covers the application of shields to types listed under MIL-W-5086A.

MIL-C-25038(USAF) Specification covers high-temperature and fire-resistant electric cable for aircraft. Cable is nickel-clad copper with insulations that will operate in an ambient temperature of 650°F. It is for use in aircraft electrical circuits which may be subject to flames.

NAS-702 Specification by National Aircraft Standards Committee covers wire which operates at 105°C. It is presently finding use as a general-purpose and hookup wire in aircraft for voltages to 2,500 volts.

NAS-703 National Aircraft Standard covers high-temperature insulated, copper hookup wire. Voltage ranges are 600 and 1,000 volts. Temperature range is to 200°C from AWG 32–8.

INSULATED AND BARE COPPER WIRE TABLES

Table 1 Table of standard annealed bare copper wire using American wire gauge (B&S)

| Gauge (AWG) or (B&S) | Diameter, inches | | | Area Circular mils | Weight Pounds per M' | Length Feet per lb. | Resistance at 68° F | | | Gauge (AWG) or (B&S) |
	Min.	Nom.	Max.				Ohms per M'	Feet per ohm	Ohms per lb.	
0000	.4554	.4600	.4646	211600.	640.5	1.561	.04901	20400.	.00007652	0000
000	.4055	.4096	.4137	167800.	507.9	1.968	.06180	16180.	.0001217	000
00	.3612	.3648	.3684	133100.	402.8	2.482	.07793	12830.	.0001935	00
0	.3217	.3249	.3281	105500.	319.5	3.130	.09827	10180.	.0003076	0
1	.2864	.2893	.2922	83690.	253.3	3.947	.1239	8070.	.0004891	1
2	.2550	.2576	.2602	66370.	200.9	4.977	.1563	6400.	.0007778	2
3	.2271	.2294	.2317	52640.	159.3	6.276	.1970	5075.	.001237	3
4	.2023	.2043	.2063	41740.	126.4	7.914	.2485	4025.	.001966	4
5	.1801	.1819	.1837	33100.	100.2	9.980	.3133	3192.	.003127	5
6	.1604	.1620	.1636	26250.	79.46	12.58	.3951	2531.	.004972	6
7	.1429	.1443	.1457	20820.	63.02	15.87	.4982	2007.	.007905	7
8	.1272	.1285	.1298	16510.	49.98	20.01	.6282	1592.	.01257	8
9	.1133	.1144	.1155	13090.	39.63	25.23	.7921	1262.	.01999	9

(Continued on next page)

175

INSULATED AND BARE COPPER WIRE TABLES (Cont.)

Table 1 Table of standard annealed bare copper wire using American wire gauge (B&S)

Gauge (AWG) or (B&S)	Diameter, inches			Area	Weight	Length	Resistance at 68° F			Gauge (AWG) or (B&S)
	Min.	Nom.	Max.	Circular mils	Pounds per M'	Feet per lb.	Ohms per M'	Feet per ohm	Ohms per lb.	
10	.1009	.1019	.1029	10380.	31.43	31.82	.9989	1001.	.03178	10
11	.08983	.09074	.09165	8234.	24.92	40.12	1.260	794.	.05053	11
12	.08000	.08081	.08162	6530.	19.77	50.59	1.588	629.6	.08035	12
13	.07124	.07196	.07268	5178.	15.68	63.80	2.003	499.3	.1278	13
14	.06344	.06408	.06472	4107.	12.43	80.44	2.525	396.0	.2032	14
15	.05650	.05707	.05764	3257.	9.858	101.4	3.184	314.0	.3230	15
16	.05031	.05082	.05133	2583.	7.818	127.9	4.016	249.0	.5136	16
17	.04481	.04526	.04571	2048.	6.200	161.3	5.064	197.5	.8167	17
18	.03990	.04030	.04070	1624.	4.917	203.4	6.385	156.5	1.299	18
19	.03553	.03589	.03625	1288.	3.899	256.5	8.051	124.2	2.065	19
20	.03164	.03196	.03228	1022.	3.092	323.4	10.15	98.5	3.283	20
21	.02818	.02846	.02874	810.1	2.452	407.8	12.80	78.11	5.221	21
22	.02510	.02535	.02560	642.4	1.945	514.2	16.14	61.95	8.301	22
23	.02234	.02257	.02280	509.5	1.542	648.4	20.36	49.13	13.20	23
24	.01990	.02010	.02030	404.0	1.223	817.7	25.67	38.96	20.99	24
25	.01770	.01790	.01810	320.4	.9699	1031.	32.37	30.90	33.37	25

n									
26	53.06	24.50	40.81	1300.	.7692	254.1	.01610	.01594	.01578
27	84.37	19.43	51.47	1639.	.6100	201.5	.01434	.01420	.01406
28	134.2	15.41	64.90	2067.	.4837	159.8	.01277	.01264	.01251
29	213.3	12.22	81.83	2607.	.3836	126.7	.01137	.01126	.01115
30	339.2	9.691	103.2	3287.	.3042	100.5	.01013	.01003	.00993
31	539.3	7.685	130.1	4145.	.2413	79.7	.009028	.008928	.008828
32	857.6	6.095	164.1	5227.	.1913	63.21	.008050	.007950	.007850
33	1364.	4.833	206.9	6591.	.1517	50.13	.007180	.007080	.006980
34	2168.	3.833	260.9	8310.	.1203	39.75	.006405	.006305	.006205
35	3448.	3.040	329.0	10480.	.09542	31.52	.005515	.005515	.005515
36	5482.	2.411	414.8	13210.	.07542	25.00	.005100	.005000	.004900
37	8717.	1.912	523.1	16660.	.06001	19.83	.004553	.004453	.004353
38	13860.	1.516	659.6	21010.	.04759	15.72	.004065	.003965	.003865
39	22040.	1.202	831.8	26500.	.03774	12.47	.003631	.003531	.003431
40	35040.	0.9534	1049.	33410.	.02993	9.888	.003245	.003145	.003045
41	55750.	.7559	1323.	42140.	.02373	7.8400	.00290	.00280	.00270
42	89120.	.5977	1673.	53270.	.01877	6.2001	.00259	.00249	.00239
43	141000.	.4753	2104.	67020.	.01492	4.9284	.00232	.00222	.00212
44	227380.	.3743	2672.	85100.	.01175	3.8809	.00207	.00197	.00187
45	356890.	.2987	3348.	106600.	.00938	3.0976	.00186	.00176	.00166
46	563900.	.2377	4207.	134040.	.00746	2.4649	.00167	.00157	.00147

INSULATED AND BARE COPPER WIRE TABLES (Cont.)

Insulated Copper Wire Table

AWG of bare copper conductor	Diameter of bare copper conductor		Insulation								
			+Diameter over insulation, inches								
			Enamel & single cotton		Enamel & single silk		Enamel & nylon		Resin & single cotton		Resin & nylon	
	Min.	Max.	Min.	Max.	Min.	Max.	Min.	Max.	Min.	Max.	Min.	Max.
4	.2023	.2063	.2110	.2173	—	—	—	—	.2110	.2173	—	—
5	.1801	.1837	.1888	.1947	—	—	—	—	.1888	.1947	—	—
6	.1604	.1636	.1690	.1745	—	—	—	—	.1690	.1745	—	—
7	.1429	.1457	.1514	.1565	—	—	—	—	.1514	.1565	—	—
8	.1272	.1298	.1356	.1404	*	*	*	*	.1356	.1404	—	—
9	.1133	.1155	.1209	.1251	*	*	*	*	.1208	.1251	—	—
10	.1009	.1029	.1075	.1114	*	*	*	*	.1075	.1114	—	—
11	.0898	.0916	.0960	.0996	*	*	*	*	.0956	.0991	—	—
12	.0800	.0816	.0861	.0895	*	*	*	*	.0857	.0890	—	—
13	.0713	.0727	.0774	.0805	*	*	*	*	.0770	.0800	—	—
14	.0635	.0647	.0696	.0725	*	*	*	*	.0692	.0720	—	—
15	.0565	.0577	.0625	.0654	.0590	.0619	.0591	.0621	.0621	.0649	.0596	.0628
16	.0503	.0513	.0562	.0589	.0527	.0554	.0528	.0556	.0558	.0584	.0533	.0563
17	.0448	.0458	.0507	.0533	.0472	.0498	.0473	.0500	.0503	.0528	.0478	.0507

18	.0455	.0428	.0476	.0453	.0448	.0423	.0446	.0422	.0481	.0457	.0407	.0399
19	.0411	.0384	.0432	.0409	.0404	.0379	.0402	.0378	.0437	.0413	.0363	.0355
20	.0370	.0345	.0391	.0370	.0363	.0340	.0361	.0339	.0396	.0374	.0323	.0317
21	.0335	.0310	.0356	.0335	.0328	.0305	.0326	.0304	.0361	.0339	.0288	.0282
22	.0302	.0278	.0323	.0303	.0295	.0273	.0293	.0272	.0323	.0303	.0256	.0250
23	.0273	.0251	.0294	.0276	.0266	.0246	.0264	.0245	.0294	.0276	.0228	.0224
24	.0247	.0226	.0268	.0251	.0240	.0221	.0238	.0220	.0268	.0251	.0203	.0199
25	.0224	.0204	.0240	.0224	.0217	.0199	.0215	.0198	.0240	.0224	.0181	.0177
26	.0203	.0183	.0219	.0203	.0196	.0178	.0194	.0177	.0219	.0203	.0161	.0157
27	.0185	.0167	.0201	.0187	.0178	.0162	.0176	.0161	.0201	.0187	.0143	.0141
28	.0168	.0150	.0184	.0170	.0161	.0145	.0159	.0144	.0184	.0170	.0127	.0125
29	.0155	.0137	.0171	.0157	.0148	.0132	.0146	.0131	.0171	.0157	.0114	.0112
30	.0141	.0123	.0157	.0143	.0134	.0118	.0132	.0117	.0157	.0143	.0101	.0099
31	.0130	.0112	.0146	.0132	.0122	.0107	.0120	.0106	.0145	.0132	.0090	.0088
32	.0120	.0103	.0136	.0123	.0113	.0098	.0111	.0097	.0136	.0123	.0081	.0079
33	.0111	.0094	.0127	.0114	.0103	.0088	.0101	.0087	.0126	.0113	.0072	.0070
34	.0102	.0086	.0118	.0106	.0094	.0080	.0092	.0079	.0117	.0105	.0064	.0062
35	.0094	.0078	.0110	.0098	.0086	.0072	.0084	.0071	.0109	.0097	.0057	.0055
36	.0088	.0072	.0099	.0088	.0080	.0066	.0078	.0065	.0101	.0090	.0051	.0049
37	.0082	.0067	.0093	.0083	.0074	.0060	.0072	.0059	.0095	.0084	.0046	.0044
38	.0076	.0061	.0087	.0077	.0069	.0055	.0067	.0054	.0090	.0079	.0041	.0039
39	.0071	.0056	.0082	.0072	.0063	.0049	.0061	.0048	.0084	.0073	.0036	.0034
40	.0067	.0052	.0078	.0068	.0059	.0045	.0057	.0044	.0080	.0069	.0032	.0030

*Special construction—not standardized.

SCREWS, BOLTS, NUTS AND WASHERS

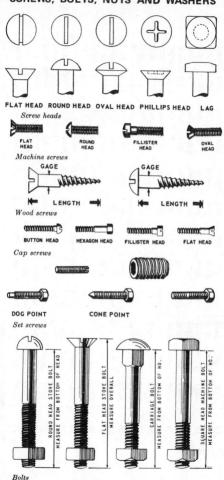

FLAT HEAD   **ROUND HEAD**   **OVAL HEAD**   **PHILLIPS HEAD**   **LAG**

Screw heads

FLAT HEAD   ROUND HEAD   FILLISTER HEAD   OVAL HEAD

Machine screws

GAGE   LENGTH   GAGE   LENGTH

Wood screws

BUTTON HEAD   HEXAGON HEAD   FILLISTER HEAD   FLAT HEAD

Cap screws

DOG POINT   CONE POINT

Set screws

ROUND HEAD STOVE BOLT
MEASURE FROM BOTTOM OF HEAD

FLAT HEAD STOVE BOLT
MEASURE OVERALL

CARRIAGE BOLT
MEASURE FROM BOTTOM OF HD.

SQUARE HEAD MACHINE BOLT
MEASURE FROM BOTTOM OF HD.

Bolts

SQUARE   HEXAGONAL   JAM   CASTELLATED

WING   CAP   THUMB   STOP

Nuts

FLAT   SPLIT LOCK   SHAKE PROOF

Washers

Studs

RIVETS AND SELF-TAPPING SCREWS

CONE HEAD | ROUND TOP COUNTERSUNK HEAD | BUTTON HEAD | TRUSS HEAD | PAN HEAD | FLAT TOP COUNTERSUNK HEAD | FLAT HEAD | TINNERS' RIVET

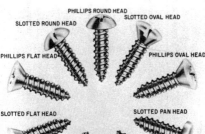

PHILLIPS ROUND HEAD

SLOTTED ROUND HEAD   SLOTTED OVAL HEAD

PHILLIPS FLAT HEAD   PHILLIPS OVAL HEAD

SLOTTED FLAT HEAD   SLOTTED PAN HEAD

PHILLIPS TRUSS HEAD   PHILLIPS PAN HEAD

SLOTTED TRUSS HEAD

TRANSDUCER TYPES

MECHANICAL	THERMAL	POWER— ELECTRICAL	LIGHT— RADIATION	MAGNETIC	CHEMICAL— PHYSICAL
Length Thickness Width	*Temperature* Bolometer Pyrometer Thermocouple Thermistor	*Current* Conductance Thermal conversion	*Optical* Photo devices	*Induction* Search coils	*Moisture—Humidity* Hygrometer Coaxial capacitor tube Thermistors
Displacement Strain R, L, C variations		*Voltage* Probes Electrode systems	*Ionization* Chambers and counters	*Moving charge* Rotating disk Magnetometer Hall effect devices	*Acidity* pH meters
Level R, L, C variations	*Heat Flow* All above in dynamic circuits	*Charge* Electrometers Vibrating capacitor	*Thermionic* Gauges	*Permeability* Search coils Flux detectors	*Molecular—gas composition* Spectrophotometers Absorbtion cell Thermal conductivity Hot platinum wire

Pressure
RLC $\Big\}$variations
diode

Flow
Viscosity

Acceleration
Piezoelectric
crystal

Velocity
Tachometer

Space charge
Ion collector

Field
Charged plate

Luminescence
Scintillation
counters

183

TRANSDUCER CHARACTERISTICS

Type	Physical transfer	Stimulus and range	Sensitivity	Dynamic range	Impedance	Notes
Capacitive	Level proximity	A-c variable with capacitance	0–10 mmf	RF communication frequencies	Megohms	
Resistive	Thermal	−200 to 100°C	.5 to 5% per degree	Low communication frequencies	10 to 20,000 ohms	Using thermistors, barretters
Potentiometer	Angular displacement	.2° upward	.2%	Wide—limited by circuit	500 to 10,000 ohms	
Differential transformer	Linear	Displacement, inches	Fractional volts/inch	Dependent on power frequency	10 to 100 ohms	

Device	Measured quantity	Range	Sensitivity	Frequency	Impedance	Notes
Synchro-resolvers	Angular displacement	0 to 360°	2 to 10 minutes	60 or 400 cycles	10 to 100 ohms	
Strain gauge	Force	Distortion—stretch, bulge	1–400 gauge factor	Low audio	10 to 1000 ohms	High sensitivity in solid state types
Pressure cells	Force	Microinches				Carbon pile Semiconductors
Electrolytic cell	Angular tilt	0–20°				

GLOSSARY OF MICROWAVE TERMS

Attenuation. Decrease in magnitude of current, voltage, or power of a signal in transmission between points.

Attenuation constant. For a travelling plane wave of a given frequency, the rate of exponential decrease of the amplitude of a field component (or of the voltage or current) in the direction of propagation. Expressed in Nepers or db per unit length.

Attenuator, flap. A device designed to introduce attenuation into a waveguide circuit by means of a resistive material moved into the guide.

Attenuator, rotary vane. A device designed to introduce attenuation into a waveguide circuit by means of varying the angular position of a resistive material in the guide.

Backward-wave tube. A travelling-wave tube in which the electrons travel in a direction opposite to that in which the wave is propagated. A microwave oscillator.

Barretter. A metallic resistor with a positive temperature coefficient of resistivity. Used for detection and power-level measurements.

Bend, E-plane. A bend in a waveguide in the plane of the electric field. ("Easy" bend.)

Bend, H-plane. A bend in a waveguide in the plane of the magnetic field. ("Hard" bend.)

Bolometer. A barretter, a thermistor, or any other device utilizing the temperature coefficient of resistivity of some resistance element.

Choke joint. A type of joint for connecting two sections of waveguide. It is so arranged that there is efficient energy transfer without the necessity of an electrical contact at the insides of the guide.

Coaxial line. A transmission line in which one conductor completely surrounds the other, the two being coaxial and separated by a continuous solid dielectric or by dielectric spacers. Such a line is characterized by no external field and by having no susceptibility to external fields from other sources.

Coupler, directional. A device consisting of two transmission lines coupled together in such a way that a wave travelling in one line in one direction excites a wave in the other guide; ideally, in one direction only.

Coupler, forward. A directional coupler used to sample incident power.

Coupler, reverse. A directional coupler used to sample reflected power.

Coupling coefficient. A ratio between the power entering the main arm of a directional coupler in one direction to the power coupled into the auxiliary arm in the same direction.

Cut-off frequency. The lowest frequency at which lossless waveguide will propagate energy in some particular mode without attenuation.

Cut-off wavelength. The longest wavelength at which lossless waveguide will propagate energy in some particular mode without attenuation.

Demodulator. A device whose output voltage is proportional to the square of its input voltage (i.e., input power).

Detector. An element which reproduces the modulation of an RF wave, usually a semiconductor crystal. Barretters are sometimes used to detect low-frequency modulation.

Directivity. The ratio of (1) power flowing out of the auxiliary arm of a directional coupler when power is flowing in the forward direction in the main arm, to (2) power flowing out of the auxiliary arm of the coupler when power is flowing in the reverse direction in the main arm (both forward and reverse powers in the main arm being equal in magnitude).

Directivity signal. A spurious signal present in the output of a coupler because the directivity of the coupler is not infinite.

Efficiency, bolometer mount. The percentage of net applied power that is absorbed by the RF termination.

(Continued on next page)

MICROWAVE TERMS (cont'd.)

EHF. Extremely high frequency. The band of frequencies between 30,000 mc (30 gc) and 300,000 mc (300 gc).

E-H tee. A junction composed of a combination of E and H plane tee junctions having a common point of intersection with the main guide.

E-H tuner. An E-H tee used for impedance transformation, having two arms terminated in adjustable plungers.

Gigacycle. 10^9 cycles (formerly kilomegacycle). Common term for expressing microwave frequencies.

Guide wavelength. The length of waveguide corresponding to one cycle of variation in the axial (transmitted) direction.

Impedance, characteristic (of a rectangular waveguide). For the dominant TE_{10} mode of a lossless rectangular waveguide at a frequency above the cut-off frequency, the ratio of the square of the rms voltage between midpoints of the two conductor faces normal to the electric vector, and the total power flowing when the guide is match-terminated.

Impedance, characteristic (of a two-conductor transmission line). For a travelling, transverse electromagnetic wave, the ratio of the complex voltage between the conductors to the complex current on the conductors.

Impedance, normalized. Any impedance of a system divided by its characteristic impedance.

Incident power or signal. Power flowing from the generator to the load.

Iris. In a waveguide, a conducting plate or plates, of small thickness compared to a wavelength, occupying a part of the cross section of the waveguide. When only a single mode can be supported, an iris acts substantially as a shunt admittance.

Isolator ferrite. A microwave device which allows RF energy to pass through in one direction with very little loss while RF power in the reverse direction is absorbed.

Junction, hybrid. A waveguide arrangement with four branches which, when branches are properly terminated, has the property that energy can be transferred from any one branch into only two of the remaining three. In common usage this energy is equally divided between the two branches.

Magnetron. A high-power microwave oscillator tube with a fixed or limited frequency range. Frequency, efficiency, and power depend on magnetic field strength and anode voltage.

MASER (Microwave Amplification by Stimulated Emission of Radiation). A low-noise, microwave amplifier utilizing a change in energy level of a material to obtain signal amplification. Common materials are gases (ammonia) and crystals (ruby).

Matched termination (waveguide). A termination producing no reflected wave at any transverse section of the waveguide.

Microstrip. A microwave-transmission component utilizing a single conductor supported above a ground plane.

Microwave region. That portion of the electromagnetic spectrum lying between the far infrared and conventional RF portion. Commonly regarded as extending from 1,000 megacycles (30 cm) to 300,000 megacycles (1 mm).

Millimeter waves. The band of frequencies having wavelengths shorter than 1 cm (above 30,000 mc)

Mismatch loss (reflection loss). The ratio, expressed in db, of the incident power to the transmitted power at a discontinuity. A measure of the loss caused by reflection.

Mode (of transmission propagation). A form of propagation of guided waves that is characterized by a particular field pattern in a plane transverse to the direction of propagation. The field pattern is independent of the position along the axis of the waveguide and, for uniconductor waveguide, independent of frequency.

(Continued on next page)

MICROWAVE TERMS (cont'd.)

Noise figure. A figure of merit for microwave amplifiers. A ratio in db between actual output noise power, and the output noise power which would come from a noiseless amplifier with identical gain and bandwidth.

Parametric amplifier (MAVAR—Mixer Amplification by Variable Reactance). A microwave amplifier utilizing the nonlinearity of a reactance element to obtain amplification. A low-noise amplifier.

Propagation constant. A transmission characteristic of a line which indicates the effect of the line on the wave being transmitted along the line. It is a complex quantity having a real term, the attenuation constant, and an imaginary term, the phase a constant.

Rat race (hybrid ring). A hybrid junction which consists of a re-entrant line (waveguide) of proper electrical length to sustain standing waves, to which four side arms are connected. Commonly used as an equal power divider.

Reflected power or signal. Power flowing from the load back to the generator.

Reflection coefficient. A numerical ratio between the reflected voltage and the incident voltage.

Reflectometer. A microwave system arranged to measure the incident and reflected voltages and to indicate their ratio (swr).

Reflex klystron. A low-power microwave oscillator tube which depends primarily on the physical size of a cavity resonator for its frequency. Normally has a wider frequency range than a magnetron.

Reike diagram. A polar-coordinate load diagram for microwave oscillators, particularly klystrons and magnetrons.

Return loss. The ratio, expressed in db, between the power incident upon a discontinuity and the power reflected from the discontinuity. (The number of db reflected power is down from incident power.)

Rotator. In waveguides, a means of rotating the plane of polarization. In a rectangular waveguide, rotation is accomplished simply by twisting the guide itself.

SHF. Super high frequency. The band of frequencies between 3,000 and 30,000 mc.

Slotted section. A length of waveguide in the wall of which is cut a nonradiating slot used for standing wave measurements.

Smith Diagram or Chart. A diagram with polar coordinates; developed to aid in the solution of transmission-line and waveguide problems.

Thermistor. A resistance element made of a semiconducting material which exhibits a high negative temperature coefficient of resistivity.

Travelling-wave tube. A broadband, microwave tube which depends for its characteristics upon the interaction between the field of a wave propagated along a waveguide and the beam of electrons travelling with the wave. A microwave amplifier.

Tuning screw (slide-screw tuner). A screw or probe inserted into the top or bottom of a waveguide (parallel to the E field) to develop susceptance, the magnitude and sign of which is controlled by the depth of penetration of the screw.

UHF. Ultra high frequency, the band of frequencies between 300 and 3,000 mc.

VHF. Very high frequency, the band of frequencies between 30 and 300 mc.

Voltage standing-wave ratio (*VSWR* or *SWR*). The measured ratio of the field strength at a voltage maximum to that at an adjacent minimum.

Wave circuits, slow. A microwave circuit designed to have a phase velocity considerably below the speed of light. The general application for such waves is in travelling-wave tubes.

Wave, dominant. The guided wave having the lowest cut-off frequency. It is the only wave which will carry energy when the excitation is between the lowest cut-off frequency and the next higher frequency of a waveguide.

(Continued on next page)

MICROWAVE TERMS (Cont.)

Waveguide phase shifter. A device for adjusting the phase of a particular field component at the output of the device relative to the phase of that field component at the input.

Waveguide tee. A junction used for the purpose of connecting a branch section of a waveguide in series with or parallel with the main transmission line.

Waveguide tuner. An adjustable device added to a waveguide for the purpose of an impedance transformation.

Waveguide wavelength. For a travelling plane wave at a given frequency, the distance along the waveguide between points at which a field component (or the voltage or current) differs in phase by 2π radians.

Wave, phase velocity. The velocity with which a point of constant phase is propagated in a progressive sinusoidal wave.

Wave, group velocity. The velocity with which the envelope of a group of waves of neighboring frequencies travels in a medium; usually identified with the velocity of energy propagation.

Wave, transverse electric (TE wave). In a homogeneous isotropic medium, an electromagnetic wave in which the electric field vectors are everywhere perpendicular to the direction of propagation.

Wave, transverse electromagnetic (TEM wave). In a homogeneous isotropic medium, an electromagnetic wave in which both the electric and magnetic field vectors are everywhere perpendicular to the direction of propagation. Generally dominant mode of coaxial lines.

Wave, transverse magnetic (TM wave). In a homogeneous istropic medium, an electromagnetic wave in which the magnetic field vector is everywhere perpendicular to the direction of propagation.

Wave, TE_{mn} (in rectangular waveguide). In a hollow, rectangular, metal cylinder, the transverse electric wave for which m is the number of half-period variations of the electric field along the longer transverse dimension, and n is the number of half-period variations of the magnetic field along the shorter transverse dimensions.

Wave, TM_{mn} (in rectangular waveguide). In a hollow, rectangular, metal cylinder, the transverse magnetic wave for which m is the number of half-period variations of the magnetic field along the longer transverse dimension, and n is the number of half-period variations of the magnetic fields along the shorter transverse dimensions.

Wavemeter, absorption. A device which utilizes the characteristics of a resonator, which cause it to absorb maximum energy at its resonant frequency when loosely coupled to a source.

TYPICAL VSWR MEASURING SET-UP USING THE SLOTTED LINE

The various elements of a typical slotted-line measuring set-up are shown in Fig. 1, demonstrating in this instance the measurement of a newly designed or un-

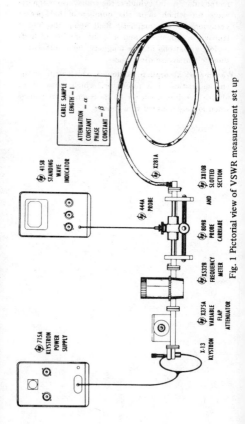

CABLE SAMPLE
LENGTH = l
ATTENUATION
CONSTANT = α
PHASE
CONSTANT = β

(hp) 415B STANDING WAVE INDICATOR

(hp) 444A PROBE

(hp) X281A

(hp) X810B SLOTTED SECTION

(hp) 809B PROBE CARRIAGE

(hp) X532B FREQUENCY METER

(hp) X375A VARIABLE FLAP ATTENUATOR

(hp) X-13 KLYSTRON

(hp) 715A KLYSTRON POWER SUPPLY

Fig. 1 Pictorial view of VSWR measurement set up

measured piece of cable. The pieces of test gear are itemized and described as follows:

1. *The klystron power supply* delivers adjustable supply and modulating voltages to the oscillator tube.

2. *Klystron oscillator tube and shield mount.* The X band carrier signals generated by the klystron serve as the input power necessary for measured indications.

3. *A level set attenuator* delivers variable amounts of power to the rest of the line for steps in measuring.

4. By using a *reaction frequency meter*, dips in the VSWR amplifier indicate adjustment frequency point coincidence with the oscillator frequency. Desired frequency can then be set.

5. *The slotted line* provides the means physically to position the broadband probe.

6. *The broadband probe* provides quantities in terms of dimension and amplitude indicating wave propagation conditions throughout the bench set-up.

7. *The standing-wave amplifier and output indicator* provides detection and amplification of the probe's output.

8. *The unknown section* is the subject of measurement.

Figure 2 shows the picture set-up transformed to block-diagrammatic form when arranged to measure a connector, using a variable short circuit for the termination element.

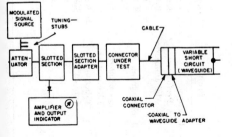

Fig. 2 Frequency variation test set-up

DATA DISPLAY DEVICES

Printed	Optomechanical	Electro-optical	
1. *Tape*	1. *Light-generated*	1. *Light-generated*	3. *Character-generated*
Perforated	Lamp arrays	Luminescent panels	Vidiac
Alphanumeric	Projected:	CRT:	Iconorama
	Image	Flexowriter	Sprendid
2. *Strip printers*	Front	Friden	Scanalog
	Opaque	Datacom	Transdata
3. *Page printers*	Tabular panels	Photorectifier panels	Randid
Electrostatic			Curveline
Fiber optics	2. *Mechanical*	2. *Light transfer*	
Impact	Message:	Register tubes	4. *Image light valves*
	Wheels	Fiber optical	5. *Dark trace*
	Plates	Edge lighting	
	Belts	Stroboscopic	
		Light interference	
		Binaview	

Recorders	Electron-optical	Projection	Special
1. *Direct writing* Single } Styli Multi }	1. *Cathode-ray tube* Typotron Tonotron Charactron	1. *Film display*	1. *Three-dimensional*
2. *Galvanometer* Direct Light beam	TV bright converter Videograph Falling raster	2. *Microfilm*	2. *Slow-scan TV*
3. *X-Y recorders*	2. *Scan-converters* Intec	3. *Readers*	3. *Facsimile*
4. *Potentiometers*	Request translator Reiterator	4. *Stereoviewer*	4. *Graph displays* Lumigraph D/A telemetry Dynatronic Quick-Look
5. *Event*	Pressar Infotracer		
6. *Sweep*	Photode-coder		
7. *Oscillograph*			

DIGITAL VOLTMETER CHARACTERISTICS

TYPE	ACCURACY Best	ACCURACY Usual	FREQUENCY RANGE	INPUT IMPEDANCE	WAVEFORM RESPONSE	NORMAL FULL SCALE RANGES	MINIMUM VOLTAGE READING	MINIMUM OBSERVABLE CHANGE
Digital Voltmeter with AC/DC Converter	0.1% of reading	0.1% to 0.5%	30 cps to 10 kc	10 megohms	average	10 v.-1,000 v.	1 mv.	1 mv.
Thermocouple Type Transfer Voltmeter	0.01% (at U.S. Nat'l Bureau Stds.)	0.03% to 0.05% of f.s.	20 cps-50 kc 0.1% acc. / to 10 mc 0.1% acc. / to 30 mc 0.2% acc.	30 ohms/v. to 500 ohms/v.	true rms	0.5 v.-1,500 v.	50 mv.	in microvolt region
Dynamometer	0.1% of f.s.	0.1% to 1% of f.s.	25 to 133 cps 0.1% acc. / any one freq. to 1,000 cps 0.1% acc.	20 ohms/v. to 100 ohms/v.	true rms	30 v.-1,000 v.	5 v.	0.5 v.

Moving Coil and Thermocouple	0.5% of f.s.	.5% to 5% of f.s.	to 5 kc 0.5% acc. can go to 5 mc	30 ohms/v. to 500 ohms/v.	true rms	0.5 v.-1,000 v.	0.1 v.	10 mv.
Moving Coil and Rectifier	0.5% of f.s.	2% to 5% of f.s.	to 10 kc — to 100 kc with lower accuracy	500 ohms/v. to 10,000 ohms/v.	average	1.5 v.-1,000 v.	0.1 v.	20 mv.
Moving Iron Vane	0.1% of f.s.	0.5% to 5% of f.s.	25-125 cps to 450 cps 1% acc.	9 ohms/v. to 60 ohms/v.	true rms	5 v.-.750 v.	1 v.	50 mv.
Electrostatic	1% of f.s.	1% of f.s.	to 100 mc	10 pf. on 60 kw. range, 225 pf. on 120 v. range. 3×10^{15} ohms leakage	true rms	120 v.-100 kw.	30 v.	1 v.
VTVM	1% of f.s.	2% to 5% of f.s.	20 cps to 10 mc — to 1 cps and to 700 mc (special)	1 megohm to 10 megohms	average (some approximate rms response for certain waveforms)	1 mv.-300 v.	20 μv.	20 μV.

THERMISTOR TYPES, SIZES, SPECIFICATIONS

Parameter	Glass coated beads	Bead in glass probes or rods	Bead in glass* enclosed envelopes	Discs	Rods	Washers
Diameter (in.)	.005 to .088	.020 to .100	.09 to .50	.05 to 1.00	.05 to .625	.2 to 1.0
Length or thickness (in.)	.088 to .135	.125 to 2.0	.75 to 3.0	.010 to .375	.3 to 3.125	.030 to .375
R_T at 25°C (ohms)	1Ω to 20M	1Ω to 20M	100Ω to 20M	4Ω to 1M	2Ω to 1M	5Ω to 1.5K
α_T at 25°C (%/°C)	2.0 to 5.5	2.0 to 5.5	2.0 to 5.5	1.35 to 5.25	1.35 to 6.50	1.35 to 5.25
δ at 25°C (mw/°C) still air (heat sink for washers)	.045 to .800	0.2 to 1.0	.004 to 1.5	.5 to 35.0	3 to 80	0.1 to 1.0
τ at 25°C (sec.) still air (heat sink for washers)	0.1 to 3.0	2 to 25	—	5 to 400	15 to 800	—

T_{max} (°C)	150 to 200	150 to 300	150 to 300	10 to 125	100 to 125	100 to 125	100 to 125
P_{max} (watts)	.012 to .050	.030 to .100	.001 to .050	.050 to 3.50	0.3 to 8.0	0.3 to 8.0	0.25 to 3.50
Stability (%/yr.)	0.1 to 0.2	0.05 to 0.20	0.05 to 0.20	0.2 to 2.5	0.2 to 2.5	0.2 to 2.5	0.2 to 2.5

*Bead in glass bulbs, etc.

Notes:

1. The *Maximum Operating Temperature* is "the maximum body temperature at which a thermistor will operate for an extended period of time with acceptable stability of its characteristics." This temperature is the result of external heating only.

2. Stability figures are based on changes in zero-power resistance for exposures up to the maximum temperature. In general, the low figure is obtained at temperatures which are somewhat below the maximum temperature and the high figures are for continuous operation at the maximum temperature.

3. The maximum power for metallized surface contact types is taken as the power required to raise the temperature of a nominal unit to its maximum operating temperature in still air at an ambient of 25°C. In all cases, the maximum power should be derated linearly to zero at 125°C for metallized surface contact types and 300°C for bead types.

THERMISTOR MEASURING CIRCUITS

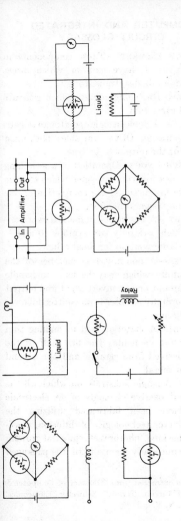

COMPUTER AND INTEGRATED
CIRCUIT CIRCUIT GLOSSARY

ACTIVE ELEMENTS—Those components in a circuit which have gain or which direct current flow: diodes, transistors.

ADDER—Switching circuits which generate sum and carry bits.

"AND"—A Boolean logic operator analogous to multiplication. Of two variables, both must be true for the output to be true.

ASYNCHRONOUS—Operation of a switching network by a free-running signal which triggers successive instructions; the completion of one instruction triggers the next.

BINARY—A system of numerical representation which uses only two symbols, 0 and 1.

BIT—Abbreviation for binary digit.

BUFFER—A non-inverting member of the digital family which may be used to handle a large fan-out or to convert input and output levels. Normally a buffer is an emitter-follower type of circuit.

CERMET—A material used in making thin film resistive elements. The first half of the term is derived from ceramic and the second half from metal.

CHIP—A single substrate on which all the active and passive elements of an electronic circuit have been fabricated utilizing the semiconductor technologies of diffusion, passivation, masking, photoresist, epitaxial growth. A chip is not ready for use until it is packaged

---

*Terms extracted from "Glossary of Computer & Integrated Circuit Terms," Schweber Electronics, Westbury, N. Y.

and provided with terminals for connection to the outside world. Also called a die.

CLEAR—To restore a memory or storage device to a "standard" state, usually the "zero" state. Also called Reset.

CLOCK—A pulse generator which controls the timing of switching circuits and memory states, and equals the speed at which the major portion of the computer operates.

CML: CURRENT-MODE LOGIC—Operates in the unsaturated mode as distinguished from all the other forms which operate in the saturated mode.

COUNTER—(a) A device capable of changing states in a specified sequence upon receiving appropriate input signals; (b) a circuit which provides an output pulse or other indication after receiving a specified number of input pulses. (Specific counters follow.)

COUNTER, BINARY—A flip-flop having a single input. Each time a pulse appears at the input, the flip-flop changes state; called a "T" flip-flop.

COUNTER, RING—A loop or circuit of interconnected flip-flops so arranged that only one is "on" at any given time and that, as input signals are received, the position of the "on" state moves in sequence from one flip-flop to another around the loop.

DCTL: DIRECT-COUPLED TRANSISTOR LOGIC —Logic is performed by transistors.

DECIMAL—A system of numerical representation which uses ten symbols 0, 1, 2, 3, . . . 9.

DELAY—Undesirable delay effects are caused by rise time and fall time which reduces

(*Continued on next page*)

COMPUTER AND INTEGRATED
CIRCUIT GLOSSARY (Cont.)

circuit speed, but intentional delay units may be used to prevent inputs from changing while clock pulses are present. The delay time is always less than the clock pulse interval.

DIFFUSION—A thermal process which introduces tiny amounts of impurities into the base material. A difficult process in solids though quite easy in fluids. Just drop a bit of coloring matter in a glass of water and the color will very gradually distribute itself throughout the water.

DIGITAL CIRCUIT—A circuit which operates like a switch, that is, it is either "on" or "off."

DISCRETE—Electronic circuits built of separate, finished components.

DTL: DIODE-TRANSISTOR LOGIC—Logic is performed by diodes. The transistor acts as an amplifier and the output is inverted.

EPITAXIAL GROWTH—A chemical reaction in which silicon is precipitated from a gaseous solution and grows upon the surface of a silicon wafer present in the gaseous solution.

EXCLUSIVE "OR"—The output is true if either of two variables is true, but not if both are true.

FALL TIME—A measure of the time required for a circuit to change its output from a high level (1) to a low level (0).

FAN-IN—The number of inputs available on a gate.

FAN-OUT—The number of gates that a given gate can drive. The term is applicable only within a given logic family.

FEB—Acronym for functional electronic block. Another name for a monolithic integrated circuit.

FLIP-FLOP—An electronic circuit having two stable states, and having the ability to change from one state to the other upon the application of a signal in a specified manner. See specific types below.

FLIP-FLOP, "D"—D stands for delay. A flip-flop whose output is a function of the input which appeared one pulse earlier, that is, if a 1 appears at its input, the output a pulse later will be a 1.

FLIP-FLOP, "J-K"—A flip-flop having two inputs designated J and K. At the application of a clock pulse, a 1 on the J input will set the flip-flop to the 1 or "on" state; a 1 on the K input will reset it to the 0 or "off" state, and 1's simultaneously on both inputs will cause it to change state regardless of what state it had been in.

FLIP-FLOP, "R-S"—A flip-flop having two inputs designated R and S. At the application of a clock pulse, a 1 on the S input will set the flip-flop to the 1 or "on" state, and a 1 on the R input will reset it to the 0 or "off" state. It is assumed that 1's will never appear simultaneously at both inputs.

FLIP-FLOP, "R-S-T"—A flip-flop having three inputs, R, S, and T. The R and S inputs produce states as described for the R-S flip-flop above; the T causes the flip-flop to change states.

FLIP-FLOP, "T"—A flip-flop having only

(*Continued on next page*)

one input. A pulse appearing on the input will cause the flip-flop to change states.

GATE—A circuit having two or more inputs and one output, the output depending upon the combination of logic signals at the inputs. There are four gates, called: AND, OR, NAND, NOR. The definitions below assume positive logic is used.

GATE, AND—All inputs must have 1-state signals to produce a 1-state output.

GATE, NAND—All inputs must have 1-state signals to produce a 0-state output.

GATE, NOR—Any one input or more having a 1-state signal will yield a 0-state output.

GATE, OR—Any one input or more having a 1-state signal is sufficient to produce a 1-state output.

GIGO—An acronym used to describe a computer whose operation is suspect. (Garbage in, garbage out.)

HALF-SHIFT REGISTER—Another name for flip-flop.

HYBRID—A method of manufacturing integrated circuits by using a combination of the monolithic and thin film methods.

INTEGRATED CIRCUIT—The Electronic Industries Association defines the semiconductor integrated circuit as "the physical realization of a number of electrical elements inseparably associated on or within a continuous body of semiconductor material to perform the functions of a circuit."

INVERTER—The output is always in the opposite logic state of the input. Also called a NOT circuit.

LINEAR CIRCUIT—A circuit whose output is an amplified version of its input, or whose output is a predetermined variation of its input.

MONOBRID—A method of manufacturing integrated circuits by using more than one monolithic chip within the same package.

"NOT"—A Boolean logic operator indicating negation. A variable designated "not" will be the opposite of its "and" or "or" function. A switching function for only one variable.

"OR"—A Boolean operator analogous to addition. (except that two truths will only add up to one truth). Of two variables, only one need be true for the output to be true.

PARALLEL OPERATION—Pertaining to the manipulation of information within computer circuitry in which the digits of a word are transmitted simultaneously on separate lines. It is faster than serial operation, but requires more equipment.

PASSIVE ELEMENTS—Those components in a circuit which have no gain characteristics, capacitors, resistors, or inductors.

POSITIVE LOGIC—The more positive voltage (or current level) represents the 1-state; the less positive level represents the 0-state.

PROPAGATION DELAY—A measure of the time required for a change in logic level to propagate through a chain of circuit elements.

RCTL: RESISTOR-CAPACITOR TRANSISTOR LOGIC—Same as RTL except that capacitors are used to enhance switching speed.

(*Continued on next page*)

REGISTER—A device used to store a certain number of digits within the computer circuitry, often one word. Certain registers may also include provisions for shifting, circulating, or other operations.

RTL: RESISTOR-TRANSISTOR LOGIC—Logic is performed by resistors. The transistor produces an inverted output from any positive input.

SERIAL OPERATION—Pertaining to the manipulation of information within computer circuitry, in which the digits of a word are transmitted one at a time along a single line. Though slower than parallel operation, its circuitry is considerably less complex.

SKEWING—Refers to time delay or offset between any two signals.

SKEWING RATE—Refers to rate at which output can be driven from limit to limit over the dynamic range.

SYNCHRONOUS—Operation of a switching network by a clock pulse generator. It is slower and more critical than asynchronous timing but requires less and simpler circuitry.

THIN FILM—A method of manufacturing integrated circuits by depositing thin layers of materials to perform electrical functions. Usually only passive elements are made this way.

TTL: TRANSISTOR-TRANSISTOR LOGIC—A modification of DTL which replaces the diode cluster with a multiple-emitter transistor.

WORD—The term "word" denotes an assemblage of bits considered as an entity in a computer.